Being Interdisciplinary

Being Interdisciplinary

Adventures in urban science and beyond

Alan Wilson

First published in 2022 by
UCL Press
University College London
Gower Street
London WC1E 6BT

Available to download free: www.uclpress.co.uk

ISBN: 978-1-80008-214-4 (Hbk.)
ISBN: 978-1-80008-213-7 (Pbk.)
ISBN: 978-1-80008-212-0 (PDF)
ISBN: 978-1-80008-215-1 (epub)
ISBN: 978-1-80008-216-8 (mobi)
DOI: https:// doi.org/10.14324/111.9781800082120

Contents

Preface

My main objective in this book is to shed light on the question of how to do interdisciplinary research. Insights have emerged, for me, from decades of research and the accumulation of a toolkit, and so this is a personal account. There are two key threads: first, that (almost?) all research is essentially interdisciplinary; and second, that a systems perspective is a good starting point, one that indeed forces us to look beyond disciplines. There is also a more subtle third thread: how to identify the game-changers of the past as a basis for learning to think outside the box of convention – how to do something new, how to be ambitious. In a nutshell, how to be creative.

I will set out some general principles for doing interdisciplinary research which should be widely applicable, whatever the background of the researcher or the nature of the research challenge. I illustrate the practice of interdisciplinary research from my own experience, but I believe that this can be easily related to challenges in the experience of others. In using this material for giving talks and running seminars, I have certainly found this to be the case.

The ideas offered here have been presented to a variety of audiences, ranging from undergraduates, masters and PhD students and early-career researchers – all in a variety of disciplines – through to those with wide experience of interdisciplinary research. In relation to 'real challenges' in my own field of urban research, I have had the pleasure of speaking at public meetings with large audiences. On such occasions I have been delighted by the interest shown in the way that research can illuminate the problems in people's minds.

I am indebted to Pat Gordon-Smith of UCL Press and two anonymous reviewers for many constructive suggestions which have greatly improved the book; many thanks to Catherine Bradley for her immaculate copy-editing and to Grace Patmore for seeing the book through to production. I am also grateful to many students, friends and colleagues who have collaborated and discussed the challenges of interdisciplinary research over the years. Last but far from least, I dedicate this book with thanks to Sarah for her love and support.

Alan Wilson
Norfolk, 2021

Prologue: a research autobiography

Interdisciplinarity is a complex subject which can be approached in many ways. I have rooted this approach in my personal experience, but aim to draw on general insights which are potentially valuable to readers from a variety of backgrounds – from science via social science to the arts and, not least, the professions. As background, I have set out a brief map of this experience in this prologue.

I have been privileged to encounter a series of partly serendipitous career challenges which took me on an interdisciplinary path before the idea was fashionable. I chart this progression not as a model, but rather as an account of possibilities. This brief account builds on a variety of experiences and choices, as well as the ideas that start to put a rationale round interdisciplinary thinking. I was fortunate in my early career to have the opportunity to spend time in North America, meeting the founding fathers – they were all men – of what became my research field. As a complement to university-based research, I was a founding director of a university spin-out company; here we had to be very business-like, which taught me something about adaptation. I also found myself later venturing into other disciplines, for example being invited 'in' to help develop a new research base. Such an experience gives a different perspective on interdisciplinarity.

I was new to geography when I went to Leeds as a Professor in 1970. There was even a newspaper headline in one of the trade papers with words to the effect that 'Leeds appoints Geography Professor with no qualifications in Geography!' So how did this come about? As I reflect, it makes me realise the extent to which my career – and this will be by no means unique – has not only been shaped by serendipity, but has also become an illustration of the emergence of interdisciplinarity.

I graduated in mathematics and I wanted to work as a mathematician. I had a summer job as an undergraduate in the (then new) Rutherford Laboratory at Harwell, which led to a full-time post when I left Cambridge. I was, in civil service terms, a 'Scientific Officer' – something that I was very pleased about because I wanted to be a 'scientist'. I even put that as my profession on my new passport. It was an interesting time. I had to write a very large computer program for the analysis of bubble chamber events in experiments at CERN (which also gave me the opportunity to spend some time in Geneva). With hindsight, it was very good initial training in what was then front-line computer science.

I also realise, with several decades of hindsight, that this was an early experience of what would now be called 'data science'. Working in a team, I was given enormous responsibilities – at a level I cannot imagine being thought appropriate for a 22-year-old today. This had the advantage of teaching me how to produce things on time, difficult though the work was. It was also the early days of large, mainframe computers and I learned a lot about their enabling significance.

At Harwell I made a decision that became a characteristic of my later career – though heaven knows why I was allowed to do it. I decided to write a general program that would tackle any event thrown up by the synchrotron. The alternative – much less risky – was to write a suite of far smaller programs, each focused on particular topologies. Again with hindsight this was probably an unwise decision, though I got away with it. The moral is: go for the general if you can. All this was very much 'blue skies research' and 'impact' was never in my mind.

Within a couple of years, I began to tire of the highly competitive nature of elementary particle physics. I also wanted to work in a field where I could be more socially useful but still be a mathematician. I therefore started applying for jobs in the social sciences in universities: I still wanted to be a maths-based researcher. All the following steps in my career were serendipitous – pieces of sheer good luck.

It did not start well. I must have applied for 30 or 40 jobs and had no positive response at all. To do something different I decided, sometime in 1962, to join the Labour Party. I lived in Summertown in North Oxford, a prosperous part of the city, and there were very few members in the local ward, Summertown. Within months I had taken on the role of Ward Secretary. We selected our candidates for the May 1963 local elections, but around February they left Oxford. It then emerged that, at this short notice, there was no time to select new candidates; the rule book said that the Chairman and Secretary of the Ward would stand instead. So it came about that in May I found myself the Labour candidate for Summertown. I duly came bottom of the poll. I enjoyed it, however, and in the following year I managed to get myself selected for East Ward. This had not had a Labour councillor since 1945, but it seemed just winnable in the prevailing tide.[1] I was elected by a majority of four after four recounts. That led me into another kind of experience – three years on Oxford City Council. This was very different from conventional research, but it did add a new dimension to the idea of interdisciplinarity. It was also the beginning of my interest in cities as a prospective subject for research.

Then came a second piece of luck. I was introduced by an old school friend to a small group of economists at the Institute of Economics and Statistics in Oxford who had a research grant from the then Ministry of Transport in cost-benefit analysis. In those days – it seems strange now – social science, even economics, was largely non-quantitative. Yet here they had a very quantitative problem and needed a computer model of transport flows in cities. We struck a deal: that I would do all their maths and computing and they would teach me economics. So I changed fields by a kind of apprenticeship.

It was a terrific time. I toured the United States – where the high-profile urban modellers were – with Christopher Foster and Michael Beesley,[2] and we met people such as Britton Harris (the Penn State Study) and I. S 'Jack' Lowry (of Model of Metropolis fame). I set about trying to build the model. The huge piece of luck was in recognising that what the American engineers were doing in developing models of flows as 'gravity models' could be restated in a format based on Boltzmann and statistical mechanics rather than Newton and gravity. This generalised the methodology. The serendipity in this case was that I recognised some terms in the engineers' equations from my statistical mechanics lectures as a student, which led to the so-called 'entropy-maximising models'. I was suddenly invited to give lots of lectures and seminars and people forgot that I had this rather odd academic background.

It was a time of rapid job progression. I moved with Christopher Foster to the Ministry of Transport when he was appointed as Director General for Economic Planning and set up something called the Mathematical Advisory Unit. It grew rapidly with a model-building brief. I had been given the title of Mathematical Adviser, from which derived the name of the unit. Strictly speaking my title should have been Economic Adviser, but the civil service economists refused to accept me as such because I was not a proper economist. I was summoned to see John Moore, the Assistant Secretary responsible for what we would now call HR. He had obviously been instructed to solve the problem. 'If you are not an economist, what are you?' he asked. I replied that I was a mathematician. 'That's fine,' he said, 'we'll call you the Mathematical Adviser.'

I was in this role from 1966 to 1968, then serendipity struck again. I gave a talk on transport models in the Civil Engineering Department at UCL, and in the audience was Professor Henry Chilver. I had left the seminar and started to walk down Gower Street when he caught up with me. Chilver told me that he had just been appointed as Director of a new research centre, the Centre for Environmental Studies (CES), and 'would I like to be the Assistant Director?' My talk had, in effect, been a job interview – not the kind of thing that HR departments would allow now. And so I moved to CES and built a new team of modellers. I worked on extending what I had learned about transport models to the bigger task of building a comprehensive urban model related to the wider planning agenda – something I have worked on ever since.

This was in 1968. By the end of the 1960s, quantitative social science was all the rage. Many jobs were created as universities sought to enter the field and I had three serious approaches: one in geography at Leeds, one in economics and one in town planning. I decided, wisely as it turned out, that geography was a broad church – in a real sense, it is internally interdisciplinary and had a record of absorbing 'outsiders'. I moved to Leeds in October 1970 as Professor of Urban and Regional Geography. And so I became a geographer. Again, the experience was terrific. I enjoyed teaching. It led to long-term friendships and collaborations with a generation that is still in Leeds, or at least academia: Martin Clarke, Graham Clarke, John Stillwell, Phil Rees, Adrian MacDonald, Christine Leigh, Martyn Senior, Huw Williams and many others – a long list. Some friendships have been maintained over the years with students I met through tutorial groups. We had large research grants and could build modelling teams. Geography, in the wider sense, did prove very welcoming and it was all – or at least mostly – very congenial. Sometime in the early 1970s I found myself as Head of Department and I also started taking an interest in some university issues. That decade was mainly about research, however, and it was very productive.

By the end of the 1970s there had been the oil crisis, cuts were in the air – déjà vu – and research funding was becoming harder to get. My next big step had its origins in a race meeting at a very cold Wetherby on Boxing Day in 1983. I was with Martin Clarke in one of the bars. On the bar's TV, we were watching Borough Hill Lad, trained near Leeds by Michael Dickinson, win the King George VI Chase at Kempton. Thoughts turned to our lack of research funding. It was at that moment, I think, that we decided to investigate the possibility of commercial applications of our models. Initially we tried to 'sell' our ideas to various management consultants, anticipating that they could do the marketing for us. We had no luck, however, so we had to go it alone: a two-person, very part-time workforce.

Our first job was finding the average length of a garden path for the Post Office. Our second was to predict the usage pattern of a projected dry ski slope. We did the programming, collected the data and wrote the reports. Things started looking up when we were awarded substantial contracts by WHSmith and Toyota, as we could then start to employ people. More of this story is told in Chapter 6; it also appears in more detail in Martin Clarke's recent book.[3] What began very modestly became GMAP Ltd, with Martin serving as the Managing Director and driving its growth. At its peak, GMAP was employing 120 people and had a range of blue-chip clients. That was a kind of real geography that I was proud to be associated with. In research terms, it provided access to data that would not normally have been available to academics. The GMAP project was very much 'research on' being carried into 'research for'.

Simultaneously in the 1980s I began to be involved in university management. I became Chairman of the Board of Social and Economic Studies and Law in the university – a Dean, in modern parlance. In 1989 I was invited to become Pro-Vice-Chancellor (PVC) at a time when there was only one such post. I left the geography department – never to return, as it turned out. The then VC became Chairman of the CVCP[4] and had to spend a lot of time in London, so the PVC job was bigger than usual.

In 1991 I found myself appointed as Vice-Chancellor and embarked on that role – not without some trepidation – on 1 October. I was VC until 2004. The role proved to be challenging, exciting and demanding. It was, in Dickens' phrase, 'the best of times and the worst of times': tremendously privileged, but also with a recurring list of very difficult, sometimes unpleasant problems. I was asked by research friends why I had taken on such a job instead of continuing in research. I responded, at least half seriously, that the job was a serious social science research challenge. But the focus here is research: I somehow managed to keep my academic work going in snatches of time, but my publication rate certainly fell.

I was Vice-Chancellor for almost 13 years, then in 2004 I was scheduled to 'retire'. This, however, seemed to be increasingly unattractive. Salvation came from an unlikely source: the Department for Education and Skills (DfES) in London. I was offered the job of Director General for Higher Education and so became a civil servant for almost three years, with policy-advising and management responsibilities for universities in England. This brought me direct experience of the 'P' in 'PDA' – an idea to be explained in Chapter 1 below. Again, it was privileged and seriously interesting to be working with ministers and having a front row seat on the politics of the day.

I always knew I wanted to research again, however, so after a brief sojourn in Cambridge I returned to academic life as Professor of Urban and Regional Systems at UCL. This was another terrific experience. I worked with Mike Batty, an old modelling friend, in the Centre for Advanced Spatial Analysis, along with a group of young researchers.

Particularly exciting was the fact that my research field developed into the realms of what became known as 'complexity science', a hot topic. Research grants flowed again. This included £2.5M for a five-year grant from EPSRC[5] for a project on global dynamics. This embraced migration, trade, security and development aid: big issues to which real geography can make a significant contribution. The grant funded half a dozen new research posts and five PhD studentships.

The final step was my move to The Alan Turing Institute in the summer of 2016. The original plan was to develop a programme in urban modelling, partly on the basis that modelling was a crucial element of data science which was to some extent being neglected. However, this plan had to be put on hold as, in September, I was appointed as CEO of the Institute. This meant that I had to learn much more than anticipated about data science and AI, which delivered its own research benefits. I began to see new possibilities of urban models being embedded in 'learning machines'.

My career trajectory has thus taken me from mathematics and elementary particle physics into geography and the social sciences via economics, on to complexity science, and on again into data science and AI. Add to this elements of operational research and 'management', both in a university and as a civil servant. The most crucial moves were not planned, so I think that a 'serendipity' label is more than appropriate. As will be obvious, the different roles have always provided the opportunity to learn new skills and to respond to new challenges. I hope the following chapters may help readers who find themselves in similar new situations.

Notes

1   Many years later, when I was Vice-Chancellor in Leeds, I was at an event at which Sir John Walker, who had recently been awarded a Nobel Prize, was a guest. I introduced myself to him and he looked me up and down and said, 'I know you – I canvassed for you in Oxford in 1964'.
2   Foster and Beesley, 'Estimating the social benefit of constructing an underground railway in London'.
3   Clarke, *How Geography Changed the World, and my Small Part in it.*
4   The Committee of Vice-Chancellors and Principals.
5   EPSRC is one of the research councils that are now part of UKRI (UK Research and Innovation) – the Engineering and Physical Sciences Research Council.

Part 1
Interdisciplinary research

Chapter 1
Interdisciplinary research: a systems approach

Introduction: the organisation of this book

The foundations of this book rest on the conjecture that the most important research problems are essentially interdisciplinary. This does not exclude, of course, important research within disciplines, but such research can usually be seen as an element of something bigger. The argument is constructed in 4 parts and 10 chapters.

Part 1 introduces the systems basis of interdisciplinarity (S), the theory of how a system works (T) and the methods that can be used to represent and work with a system (M) – the STM set of building bricks. This is coupled with a framework for applications of systems research in different kinds of planning with three elements: policy – what are the 'applied' objectives (P); design – inventing alternative ways of creating plans and problem solving (D); and analysis – our understanding of the system of interest (A), the PDA framework. The two frameworks complement one another: STM is essentially the analytics basis of PDA. Such a coupling is illustrated by reference to the application of research to 'real challenges'. This introduction of key concepts in Chapter 1 is followed in Chapter 2 by an exploration of the nature of interdisciplinarity and its relation to traditional disciplines. Two key concepts are introduced here: 'requisite knowledge' and 'combinatorial evolution'.

Part 2 shifts to the practice of interdisciplinary research. It begins with consideration of 'how to start' in Chapter 3, followed by the introduction of another core concept – that of a model of a system – in Chapter 4. Research is underpinned by access to data, so Chapter 5 explores the impact of 'big data' and 'data science' before leading into machine learning and artificial intelligence (AI). The threads of the approach to practice are drawn together through a return to the topic of 'real challenges' and problem-solving in Chapter 6.

Part 3 then progresses into expanding the interdisciplinary research toolkit: how to explore the literature – especially understanding the game-changers of the past in Chapter 7. In Chapter 8 we explore the possibilities of translating ideas between domains, including the notion of superconcepts that transcend disciplines.

Part 4 explores the tools needed to manage research (and ourselves as researchers). In Chapter 9 we assess the materials and skills needed – textbooks, courses or experience, opportunities for collaboration, time management and the practice of writing. In Chapter 10 we investigate the organisation of research and its ecosystem of researchers, users and funders.

In this opening chapter, I present in turn the STM and PDA frameworks (pp. 4–6) and then illustrate their application in terms of 'real challenges', using my own experience of working on cities as an example (pp. 6–11). I broaden this range of challenges and potential interdisciplinary research problems in the exercises at the end of the chapter.

Key concepts

When I was studying physics I was introduced to the idea of a *system of interest*: defining at the outset that 'thing' you were immediately interested in studying. This is crucial for interdisciplinarity. There are three ideas to be developed from this starting point. First, it is important simply to list all the components of the system; second, to simplify as far as possible by excluding all possible extraneous elements from this list; and third, taking us beyond what the physicists were thinking at the time, to recognise the idea of a 'system' as made up of related elements; it was thus as important to understand these relationships as it was to enumerate the components. All three ideas are important. Identifying the components will also lead us into counting them; simplification is the application of Occam's Razor to the problem; and the relationships take us into the realms of interdependence and complexity. Around the time I was learning about the physicists' ideas of a 'system' something broader, 'general systems theory', was in the air, though I did not become aware of it until much later.[1]

So, always start a piece of research with a 'system of interest'. Defining the *system* then raises other questions that must be decided at the outset, albeit open to later modification: questions of scale.[2] There are three dimensions to this. First, what is the granularity at which you view your system components? Population by age: how many age groups? Or age as a continuous variable? Usually too difficult. Second, how do you treat space? Continuous with Cartesian coordinates? Or as discrete zones? If the latter, what size and shape? Third, since we will always be interested in system evolution and change, how do we treat time? Continuously? Or as a series of discrete steps? If the latter, is the focus upon one minute, one hour, one year, 10 years, or what? These choices have to be made in a way that is appropriate to the problem and also, often, in relation to the data which will be needed. (Data collectors have already made these scale decisions.)

Once the system is defined, we have to ask a question along the lines of: how does it work? Our understanding, or 'explanation', is represented by a *theory*. There may be an existing theory, whether partially or fully worked out, or there may be very little theory. Part of the research problem is then to develop the theory, possibly stated as hypotheses to be tested.

Then there is usually a third step relating to further questions. How do we represent our theory? How do we do this in such a way that it can be tested? What *methods* are available for doing this?[3]

In summary, a starting point is to define a system of interest, to articulate a theory about how it works and to find methods that enable us to represent, explore and test the theory: the *STM approach*.

This, through the reference to theory (or hypothesis) formulation and testing, establishes the science base of research. Suppose now that we want to apply our science to real-world problems or challenges. We need to take a further step which is in part an extension of the science; it is still problem-solving in relation to a particular system of interest, but has added dimensions beyond what is usually described as 'blue skies science'. The additions involve articulating objectives and inventing possible solutions.

Take a simple example: how to reduce car-generated congestion in a city. The science offers us a mathematical-computer model of transport flows. Our objective is to reduce congestion. The possible solutions range from building new roads in particular places to improving a public transport system to divert people from cars. Each possible solution can be thought of as a *plan* and the whole activity is a form of *planning*. In some cases, computer algorithms can invent plans, but more usually it is a human activity. For any plan, the new flows can be calculated using the model, along with indicators of, for example, improved traffic speeds and consumers' surplus. A cost-benefit analysis can be carried out and the plan that has the greatest rate of return or the greatest benefit-to-cost ratio chosen. In reality, of course, it is never as neat as this.[4]

How can we summarise this process? Many years ago I learned from my friend and collaborator Britton Harris that this can be thought of as *policy, design and analysis: a PDA* framework to complement the *STM* approach. His insight was that each of the three elements involved different kinds of thinking – and that it was rare to find these in one person, in one room or applied systematically to real problems. There is a further insight to be gained: the PDA framework can be applied to a problem in 'pure' science. The objective, the policy, may be simpler – to articulate a theory – but it is still important to recognise the 'design' element: the invention at the heart of the scientific process. Of course, this applies very directly to engineering, as engineers confront both a science problem and a policy and planning problem: both STM and PDA apply.

As we noted briefly, adopting a systems perspective at the outset of a piece of research forces interdisciplinarity. This can be coupled with an idea which will be developed more fully later: requisite knowledge. This is simply the knowledge that is required as the basis for a piece of research. When this question is asked about the system of interest, it will almost always demand elements from more than one discipline; these elements then combine into something new – more than the sum of the parts. There is here a fundamental lesson about effective research: it has to be interdisciplinary at the outset.

This provides a framework for approaching a subject, but we still have to choose. These decisions can be informed but are ultimately subjective. A starting point is that they should be interesting to the researcher but also important so some wider community. The topic should be ambitious but also feasible – a very difficult balance to strike.

Real challenges

Cities provide a system of interest which is complex and interesting, both as science and as science applied to real challenges. Urban research draws on social, economic and geographic disciplines and so provides a good example for our explorations. One starting point is to reflect on the well-known and very real challenges that cities face. The most obvious ones can be classified as 'wicked problems'[5] in that they have been known for a long time, usually decades, and governments of all colours have made honourable attempts to 'solve' them. Given the limited resources of most academic researchers, it could be seen as wishful thinking to take these issues on to a research agenda, but nothing ventured, nothing gained. We need here to bear in mind the PDA framework – policy, design, analysis. Good analysis will demonstrate the nature of the challenges. Policy is usually to make progress with meeting them, while the hard part is the 'design' – inventing possible solutions. The analysis comes into play again to evaluate the options. If we focus on the UK context, we can offer a sample of the issues (though most translate easily across national boundaries). Broad headings, reflecting the definition of our system of interest, might be:

1   Living in cities – people issues.
2   The economy of cities – organisations providing goods and services.
3   Urban metabolism: energy and materials flows.
4   Urban form.
5   Infrastructure.
6   Governance.[6]

The challenges have all been exacerbated by the COVID pandemic to which we return after pre-pandemic considerations – which are still with us. Consider a sample of issues within each topic.

In this example we see that since the roots lie in social, economic and geographic research, a focus on people at the outset is a good starting point. To illustrate: how do people live in cities? The challenges are well known and can be summarised. Housing: in the UK there is a current shortage and this situation will be exacerbated by population growth. Education – a critical service, upskilling for futureproofing, and yet a significant percentage of young people leave school with inadequate literacy, numeracy and work skills. Health – an uneven delivery of services. The future of work – what will happen if the much predicted 'hollowing out' occurs as middle-range jobs are automated? How will those then redundant pay their bills? A complementary perspective focuses on the economy of cities, embracing private and public sectors and the delivery of products, services and jobs (and therefore incomes).

How we live, and how the economy can function, connects strongly the greatest of contemporary challenges, climate change, along with sustainability and the feasibility – indeed the necessity – of achieving low carbon targets. This challenge is represented by the metabolism of the city in terms of energy and materials flows. However, it also connects strongly to the next three items on the list: urban form, infrastructure and governance.

Urban form raises issues such as the appropriateness of low densities in housing development and the locations of the new housing that will be necessary to meet the demands of a growing population. The continuing trends towards lower densities and the increasing average length of trips challenge our ability to meet sustainability targets. The physical embodiment of the city's form is its infrastructure. We need to relate objectives in this respect – not simply to request more public transport, for example, but to conceptualise and measure accessibilities crucial for both people and organisations, working towards transport infrastructure that underpins an effective system. Investment in utilities will also be necessary, not only to match population growth but also to respond to the sustainability agenda. In particular, counting communications and broadband as utilities, how do we secure our future in a competitive world?

To begin to meet these challenges, we need effective governance that incorporates community engagement. At what levels are planning and policy decisions best made? How do we make food, communications and utilities secure? How do we handle competing and conflicting objectives?

This brief analysis leads us to an immediate, important conclusion: these issues are highly interdependent. One important area of research is thus to chart these interdependencies and to build policies and plans that take them into account. An obvious research priority is the need for comprehensive urban models, which form the underpinning science of cities. There are some excellent examples, but they are not deployed as a core part of planning practice. Ideally, therefore, in relation to the issues sketched above, a comprehensive urban model should have enough detail in it to represent all the problems and any planning proposals should be tested by the runs of such a model. Neither of these ambitions is fulfilled in practice, and so this poses a challenge to modellers as well as those directly concerned with real-world issues. Crucially, any specification of these issues will be interdisciplinary. Bearing this in mind, let us now work down another level and pose questions about research priorities.

Living in cities means people issues. We noted earlier the housing shortage, exacerbated by population growth. The demography is itself very much worth further investigation: populations in many places are, relatively, ageing and others are being restructured by migration – both in and out. These shifts in many cases relate to work opportunities, linkages on which there has been relatively little research. We need an account, that is, a model, of where people choose to live in relation to their incomes, housing availability, affordability and prices, local environments and access to work and services – a pretty tall order for the initial analysis. There is then a planning issue that links closely to urban form: where should the new housing go? At present it is mainly located on the edges of cities, towns and villages with no obvious functional relationship to other aspects of people's lives. Related research needs be done on developers and house builders and their business models.

Education is a critical service, as we have noted. It provides upskilling for future proofing and yet in the short term has to deal with a significant percentage of pupils who leave school with inadequate literacy, numeracy and work skills. Some progress has been made in developing federations of schools to bring 'failing' schools into a more successful fold. Hard analysis remains to be done on other factors, however, particularly the impact of children's social background and whether schools' initiatives need to be extended into a wider community. Again there are examples of improvement and it should be possible to explore the relative successes of initiatives in a wide range of areas. A particular category of concern is children in care. The system here is obviously failing, as measured by the tiny percentage who progress into higher education and the high proportion of offenders who have at some time been in care.

Health is another critical service, unevenly delivered across the country. In terms of research possibilities, this sector is data rich but under-analysed – perhaps in part because of the difficulties researchers have in accessing the data. There are many research projects in the field, but it remains relatively fragmented. For example, does anyone explore an obvious question: what is the optimum size of GP surgeries in different kinds of locations? Indeed, is there a possible plan to be explored that would relocate GP surgeries into hospitals?

The economy embraces both private and public sectors, and so has to deliver products, services and jobs (and therefore incomes). Interesting research has been done on growing and declining sectors in the economy, for example. This can then be translated down to the city level and combined with the 'replicator' and 'reinventor' concepts introduced in the Centre for Cities' *Century of Cities* paper.[7] This would enable at least short-term predictions of employment change, which could also be related to the immigration issues. The major challenge for the economy is the ability to deliver employment in the context of the 'hollowing out' as middle-range jobs are automated.

We have noted what may be the biggest challenges of all: the re-working of the urban metabolism (energy and materials flows) to achieve sustainability and low-carbon futures, and the feasibility of achieving low-carbon targets. An obvious research issue here is the monitoring of current trends and analysis of past trends in relation to sustainability targets. It is likely that current trends are in the 'wrong' direction, with trips getting longer and densities decreasing. If this is the case, can we invent and test alternative futures such as shorter trips, all made by new forms of public transport or some high-density development aimed at groups who might appreciate it?

In the case of urban form, achievement of sustainability targets will make huge demands for fundamental change. Where will the necessary new housing go? Can we explore possible 'green belt' futures – for example analysing the URBED Wolfson Prize model?[8] More importantly, how can higher densities and sustainable forms be achieved, given the lock-in of present structures and market demand?

In the case of infrastructure, accessibilities are crucial for both people and organisations. Transport infrastructure and an effective system are thus correspondingly critical – at scales ranging from the neighbourhood, to the city, the region, national and global. Rural areas offer a particular research challenge: I have heard it suggested that counties could be seen as distributed cities.[9] There is an argument for some systematic research on accessibilities, a concept introduced earlier, and the ways in which they can be related to utility functions. Investment in utilities will be necessary not only to match population growth, but also to respond to the sustainability agenda. In particular, counting communications and broadband as utilities, how do we secure our future in a competitive world?

Governance: at what levels are planning and policy decisions best made? National, regional, city or neighbourhood? Or is a mixture needed? A good research question would be: how to chart subsidiarity principles to delegate downwards, from national through cities and regions to neighbourhoods while at the same time setting national-level objectives?

This is a very partial and briefly argued list, but I think it exposes the paucity of both particular and integrated research on some of the big challenges. The recognition of interdependence, as we have noted, is crucial. Consider Figure 1.1 as a broad representation of a city system.

The left-hand (grey) side of the diagram represents the demography of the city, generating individuals and households who live, work and use a variety of services. The right-hand (black) side represents the economy of the city, generating housing, employment, products and services. These categories are all linked through the (white) transport and communications subsystems. Inevitably, much research may focus on a system of interest, as per one of these boxes. An awareness of the place of these in the wider system is essential. Interdisciplinarity is crucial.

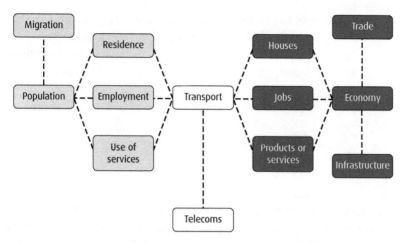

Figure 1.1 A city system.

Exercises

The following exercises are intended to help you build your own knowledge map to set up an interdisciplinary project. It would also be helpful for you to develop your own reading list in the light of the 'requisite knowledge' and 'combinatorial evolution' ideas (anticipating Chapter 2) for your own project.

1 STM

For a research domain of your own choosing:
- Define the entities that make up the system of interest and note their interrelationships. Include links to entities in other systems that provide your 'environment' – systems are almost never self-contained. In noting interrelationships, start to examine the interdependence between the elements of your system.
- Construct a mind map.[10]
- Begin to make choices about the scale at which you will describe your system:
    o detail for each kind of entity – e.g. age groups for a population
    o time
    o geographical location
- Consider how the system functions. Bearing in mind that 'theory' is the 'understanding' of your system of interest achieved in earlier research, examine the writers who have contributed and reported. Or establish

a hypothesis which you can subsequently test. Begin to articulate the concepts which are the building blocks of your understanding.

- Decide how you will describe the elements of the system in order to do the research effectively – for example, will it be quantitative or qualitative? In some circumstances a good algebraic notation along with some 'counts' – an account – makes a good starting point.

2 Real challenges – PDA

Are there any 'real challenges' associated with features of your system – for example, a housing shortage, homelessness, access to education and future-proofing skills, meeting sustainability goals – but specific to your system? There almost certainly will be. Apply the PDA framework: objectives to seek to bring about change, plans for this; inventing possible 'solutions' or plans; having the analytic capability to underpin your policy and planning ambitions.

Notes

1   For example, see von Bertalanffy, *General System Theory*.
2   Scale questions define disciplines and subdisciplines: quantum to cosmology; ethnography and psychology to social policy.
3   'Mathematics or statistics?' is a major question here.
4   Boyce and Williams, *Forecasting Urban Travel* sets out the history of this in some detail.
5   A term introduced by Rittel and Webber, 'Dilemmas in a general theory of planning', 155–69.
6   These are taken from the framework used in the Future of Cities project reports – see Government Office for Science, 2013.
7   Swinney and Thomas, 'A century of cities'.
8   Rudlin and Falk, 'Uxcester Garden City'.
9   See Herefordshire City, *Place*, Hereford Civic Society, 2015
10   For example see Buzan and Buzan, *The Mind Map Book*.

Chapter 2
Disciplines and beyond

Introduction

The construction of knowledge has evolved through a system of disciplines and we take this as a starting point (pp. 14–15). Disciplines can be defined as enablers – such as mathematics, philosophy or computer science – or in relation to the big 'real' systems – for illustration, say, the physical, the biological, the socio-economic and the environmental. We can then add the professional disciplines such as medicine and law, which of course are essentially interdisciplinary in their knowledge bases. These disciplinary areas subdivide and then subdivide further, usually through a number of increasingly specialised layers.

We then begin to explore how to create an interdisciplinary perspective and to consider how this is driven by a systems' perspective (pp. 15–18). This broad account is brought into sharp focus through the introduction of two concepts: requisite knowledge (pp. 18–20) and combinatorial evolution (pp. 18–22). The first of these drives us to accumulate everything we need to know about a system of interest for our research, capturing the essence of interdisciplinarity. The second shows that our knowledge base will always have a hierarchical structure and, intriguingly, that research breakthroughs are nearly always triggered at the lower levels, possibly in another discipline or interdisciplinary domain.

Disciplines

Disciplines are social coalitions and wield considerable power. Expertise in a discipline usually involves high levels of skill and deep learning. Conferences, journals and probably the majority of learned societies are organised as disciplines. There are even powerful coalitions of subdisciplines with even further subdivisions into factions – Tony Becher's 'academic tribes'.[1] Here, I follow the argument first put in *Knowledge Power*[2]: that it is possible to define disciplines in a systematic way, as a means towards understanding them, and then to consider how they fit into an interdisciplinary framework. It can be argued that there are (in broad terms) three kinds of discipline: those that are abstract and enabling, those defined in terms of big systems and those rooted in professions.

As examples of the first, the enablers, consider philosophy, mathematics, and computer science. Philosophy could be said to be about how to think clearly and it certainly crosses disciplines; mathematics does have a life of its own, but is particularly valuable in providing the underpinnings of many disciplines; while computer science has become the enabling discipline par excellence. As we will see in later chapters, there may now be an argument for adding 'AI and data science' as a fourth enabler.

The 'big systems' disciplines at a high level can be defined in terms of the physical, biological and the social – including the humanities in the 'social'. There are then specialisms within each – physics and chemistry, for example, within the physical sciences. Then there is much differentiation by scale – from the micro to the cosmic in physics, for example. The beginnings of one kind of interdisciplinarity can be seen with interactions across the big system boundaries and the evolution of new (sub) disciplines or coalitions such as biochemistry and biophysics.

The professional disciplines, medicine being a striking example, are already interdisciplinary, along with the addition of skills for professional practice. Consider, for example, law, engineering and planning in this context. These disciplines can usually be identified by a concept of practitioners being 'chartered' – licensed to practise.

There is an organisational challenge that holds back interdisciplinary research: universities are mostly organised in departments of traditional disciplines. They have responded to the needs of interdisciplinarity by setting up large numbers of research centres and institutes, often driven by the research councils' laudable interdisciplinary strategies. However, departments typically retain the teaching of undergraduate students and therefore the bulk of the core funding.[3] Academics are often discipline-bound by the promotion criteria and procedures in their universities which is reinforced, for example, by the importance of publications in 'top journals' rooted in disciplines.

In the next sections, we show how systems thinking and the idea of requisite knowledge drives, indeed forces, interdisciplinarity. This is reinforced by Brian Arthur's idea of combinatorial evolution which shows how ideas cross disciplines and lead to the formation of new coalitions.

Interdisciplinarity and superconcepts

Systems thinking, as introduced in Chapter 1, drives us to interdisciplinarity: we need to know *everything* about the system of interest in our research, and that means anything and everything that any relevant discipline can contribute. For almost any social science system of interest, there will be available knowledge at least from economics, geography, history, sociology, politics and psychology, plus enabling disciplines such as mathematics, statistics, computer science and philosophy; and many more.[4] This perspective points up the narrowness of disciplinary approaches. We have recognised that the professional disciplines such as medicine already have a systems focus and so in one obvious sense are interdisciplinary. But in the case of medicine the demand for in-depth knowledge has generated a host of specialisms which again produce silos and a different kind of interdisciplinary challenge, and possibly under-developed areas in research. There is the special case of medical diagnosis, for example, which is under-researched and which is interdisciplinary par excellence.

Some systems' foci are strong enough to generate new, if minor disciplines. Transport Studies is an example, though perhaps dominated by engineers and economists.[5] There is a combinatorial problem here. In terms of research challenges, there are very many ways of defining systems of interest. They are not mostly going to turn into new disciplines.

How do we build the speed and flexibility of response to take on new challenges effectively? A starting point might be the recognition of 'systems science' as an enabling discipline in its own right that should be taught in schools, colleges and universities along with, say, mathematics. This could help to develop a capability to recognise and work with transferable concepts – superconcepts – and generic problems (for which 'solutions', or at least beginnings, exist). In my *Knowledge Power* book I identified over one hundred candidate superconcepts. We illustrate these with the sample below.

Superconcepts that can cross (or transfer) between disciplines provide valuable insights into the structure of knowledge that is transdisciplinary. They are therefore part of the toolkit of the interdisciplinary researcher. A sample of these superconcepts follows.

- Systems (scales, hierarchies, etc) .
- Accounts (and conservation laws).
- Hierarchy (scales).
- Probabilities (and uncertainty).
- Entropy.
- Equilibrium (entropy, constraints, …).
- Optimisation.
- Non-linearity, dynamics (multiple equilibria, phase transitions, path dependence).
- Evolution, 'DNA', initial conditions.
- Lotka-Volterra-Richardson dynamics.

Many of these are developed in more detail in subsequent chapters, particularly in Chapter 8, but a brief sketch of this sample is given here. Each of these has a set of generic problems associated with them (though this argument was not fully developed in the *Knowledge Power* book).

The systems' concept is a key starting point, used throughout this book. Any systems of interest that come to mind allow us to assemble a tailored toolkit, partly building on superconcepts and associated generic problems, integrated with deep domain knowledge. This is the heart of the interdisciplinary challenge.

Systems entities can nearly always be accounted for – literally *counted*. In a time period, they will define a system state at the beginning and a system state at the end; entities can enter or leave the system during the period. This applies, for example, to populations, goods, money and transport flows. In each case, an account can be set out in the form of a matrix and this is usually a good starting point for model building. This is direct in demographic modelling, in input-output modelling in economics, and in transport modelling.[6]

The behaviour of most entities in a social science system is not deterministic, and therefore the idea of probability is important. The modelling task, implicitly or explicitly, is to estimate probability distributions. We often need to do this subject to various constraints, that is, prior knowledge of the system such as a total population. It then turns out that the most probable distribution consistent with any known constraints can be estimated by maximising an entropy function, or through maximum likelihood, Bayesian or random utility procedures, all of which can be shown to be equivalent in this respect – a superconcept kind of idea in itself. Which approach is chosen is likely to be a matter of background and taste. The generic problem in this case is the task of modelling a large population of entities which interact only weakly with one another. These two conditions must be satisfied. Then the method can generate, usually, good estimates of equilibrium states of the system (or in many cases, subsystem). It can also be used to estimate missing data and to estimate complete data sets from samples following model calibration based on the sample. We will return to this when we discuss Weaver's work in Chapter 7 (pp. 92–94).

Many hypotheses, or model-building tasks, involve *optimisation* and there is a considerable toolkit available for these purposes. The methods of the previous paragraph all fall into this category, for example. However, there may be direct hypotheses such as utility or profit maximisation in economics. It is then often the case that simple maximisation does not reproduce reality – because of imperfect information on the part of participants, for example. In this case, a method like entropy-maximising can offer 'optimal blurring'.[7]

The above examples focus on systems in equilibrium and, in that sense, on the fast dynamics of systems. The implicit assumption is that, after a change, there will be a rapid return to equilibrium. We can then shift to the slow dynamics – for example, in the cities case, evolving infrastructure. We are then dealing with (sub)systems that do not satisfy the 'large number of elements/weak interactions' conditions.

These systems typically represent nonlinear dynamics and a different approach is needed. Such systems have generic properties: multiple equilibria, path dependence and the possibility of phase changes – the last being abrupt transitions at critical parameter values. Examples of phase changes are the shift to supermarket food retailing in the early 1960s and ongoing gentrification in central areas of cities.

The notion of evolution is an important superconcept. In principle, we can model different kinds of evolution with the toolkit of nonlinear dynamics. We have already noted the importance of representing constraints, and therefore the initial conditions, in equilibrium modelling. These become particularly important in dynamic modelling: path dependence is the system progressing through what can be thought of as a sequence of initial conditions, at each point in time constraining further evolution.

Working with Britton Harris in the late 1970s, we evolved, on an ad hoc basis, a model to represent retail dynamics which did indeed have these properties. It was only later that I came to realise that the model equations were examples of the *Lotka-Volterra equations* from ecology. In one version of the latter, species compete for resources; in the retail case, retailers compete for consumers – and this identifies the generic nature of these model-building problems. The extent of the range of application is further illustrated by Richardson's work on the mathematics of war. It is also interesting that the work of Lotka, Volterra and Richardson all took place in the 1920s and 1930s, serving to illustrate a different point: that we should be aware of the modelling work of earlier eras for ideas for the present.[8] The path-dependent nature of these dynamic models illustrates the earlier point and accords with intuition. For example, the future development of a city depends strongly on what is present at a point in time. Path dependence, as we have seen, is a sequence of 'initial conditions' – the data at a sequence of points in time. This offers a potentially useful metaphor: that these initial conditions represent the 'DNA' of the system.

These illustrations of the nature of interdisciplinarity obviously stem from my own experience – my own intellectual toolkit that has been built over a long period. The general argument is that to be an effective contributor in interdisciplinary work it is worthwhile to build intellectual toolkits that serve particular systems of interest, a process that involves wide surveys of the possibilities in breadth as well as depth (something that is still very much needed). This leads directly into the notion of 'requisite knowledge' which is explored in the next section.

Requisite knowledge

We now work towards a law which provides a basis for interdisciplinarity. W. Ross Ashby was a psychiatrist who, through books such as *Design for a Brain*,[9] was one of the pioneers of the development of systems theory in the 1950s. A particular branch of systems theory was cybernetics – from the Greek for 'steering' – essentially the theory of the 'control' of systems. This was, and I assume is, very much a part of systems engineering and it attracted mathematicians such as Norbert Weiner.[10] For me, an enduring contribution was Ashby's Law of Requisite Variety, which is simple in concept and anticipates much of what we now call complexity science. 'Variety' is a measure of the complexity of a system and is formally defined as the number of possible 'states' of a system of interest. A coin to be tossed has two possible states – heads or tails; a machine can have billions. Suppose some system of interest has to be controlled for simplicity – a robot, say. Then the law of requisite variety asserts that the control system must have at least the same variety as the machine it is trying to control. This is intuitively obvious since any state of the machine must be matched in some way by a state of the control unit: it needs an 'if this then that' mechanism. Now suppose that the system of interest is a country and the control system is its government. It is again intuitively obvious that the government does not have the 'variety' of the country, and so its degree of control is limited. Suppose further that the government of a country is a dictatorship and wants a high degree of control. This can only be achieved by reducing the 'variety' of the country through a system of rules.

The law of requisite variety can be seen as underpinning the argument for devolution from 'central' to 'local' – a way of building 'variety' into governance. This is an idea that can be applied in many situations, for example, in universities and research councils. We begin to see how a concept that appears rooted in engineering can be applied more widely.

We can now take a bigger step and apply it to 'knowledge', specifically to the knowledge required to make progress with a research problem. The problem is now associated with a system of interest and the requisite knowledge is that which is required to make progress with the research problem. The application of the law of requisite variety can then be interpreted as relating to the specification of the toolkit of knowledge elements needed and the law asserts that it must be at least as complex as the problem. We might think of this as the 'requisite knowledge toolkit'. It seems to me that this is an important route into thinking about how to do research. What do I need to know? What do I need in my toolkit? It forces an interdisciplinary perspective at the outset.

Consider, as an example, the housing problem in the UK (see 'Real challenges' in Chapter 1 (pp. 6–11) which we elaborate here). What is the requisite knowledge which would be the basis for shifting from building the current 150,000 new houses per annum in the UK to an estimated 'need' of more than 300,000 per annum? We can get a clue from 'How to start' (Chapter 3, pp. 27–35 below). There will be policy, design and analysis elements of the toolkit. Elementary economics will tell us that builders will only build if the products can be sold, which in turn means if they can be afforded – the basic rules of supply and demand. Much of the price is determined by the price of land, so land economics is important. If land price is too high for elements of need to be met, there may be an argument for government subsidies to generate social housing. Alternatively, prices could be influenced by the cost of building. This raises questions of whether new technology could help – and this brings engineering (and international experience) very much into the toolkit. Given that there is likely to be a substantial expansion, geography kicks in – where can this number of houses be built? This is in part a question of 'where across the UK?' and in part of 'where in, or on the periphery of, particular cities'? Or should there be new 'garden cities'? All of this raises questions for the planning profession. The builders are part of a wider ecosystem, which includes land owners and the government, in relation to the regulation of land through taxation or other means. This all becomes part of the research task.

There are challenges for all of us who might want to work on this issue – academia, divided by discipline; the professions, functioning in silos; the landowners, developers and builders; and government, wanting to make progress but finding it difficult to corral the different groups into an effective unit. In this case the requisite knowledge idea and the notion of having a toolkit to assemble all the relevant elements for tackling a research problem can be deployed as a knowledge base. This sketch also shows that an important part of this is a capacity to assemble the right teams.

Combinatorial evolution

Brian Arthur introduced a new and important idea in *The Nature of Technology*[11], that of 'combinatorial evolution', which gives us a different kind of insight into interdisciplinarity. The argument, put perhaps overly briefly, is essentially this: a 'technology', such as an aeroplane, can be thought of as a system. Then we see that it is made up of a number of subsystems, which can be arranged in a hierarchy. Thus the plane has engines, engines have turbo blades and so on. The control system must sit at a high level in the hierarchy; then at lower levels we find computers. Arthur's key idea is that most innovation comes at lower levels in the hierarchy and through combinations of these innovations, hence the term 'combinatorial evolution'. The computer may have been invented to do calculations but, as with aeroplanes, now figures as the ubiquitous lynchpin of sophisticated control systems.

This provides a basis for exploring research priorities and, unsurprisingly, it forces us into an interdisciplinary perspective. Arthur is in the main concerned with hard technologies and with specifics, such as aeroplanes. However, he does remark that the economy 'is an expression of technologies' and that technological change implies structural change. He adds that: '...economic theory does not usually enter [here] … it is inhabited by historians'[12]. We can learn something here about dynamics, about economics and about interdisciplinarity. Let us focus, however, on cities.

We can certainly think of cities as technologies – and much of the smart city's agenda can be seen as low-level innovation that can then have higher-level impacts. We can also see the planning system as a 'soft technology'. What about the science of cities and of urban modelling? Arthur's argument about technology can be applied to science. Think of 'physics' as a system of laws, theories, data and experiments. Think of spelling out the hierarchy of subsystems and, historically, charting the levels at which major innovations have been delivered. Translate this more specifically to our urban agenda. If (in broad terms) modelling is the core of the science of cities and that (modelling) science is one of the underpinnings of the planning system, can we chart the hierarchy of subsystems and then think about research priorities in terms of lower-level innovation?

Suppose the top level is a working model – a transport model, a retail model or a Lowry-based comprehensive model.[13] We can represent this and three (speculative) lower levels broadly as follows.

- Level 1: working model – static or dynamic.
- Level 2 – STM:
  - System definition (entities, scales: sectoral, spatial, temporal), exogenous, and endogenous variables.
  - Hypotheses, theories.
  - Means of operationalising (statistics, mathematics, computers, software).
  - Information system (cleaned-up data, intermediate model to estimate missing data).
  - Visualisation methods.
- Level 3:
  - Explore possible hypotheses and theories for each subsystem.
  - Data processing, information system building.
  - Preliminary statistical analysis.
  - Available mathematics for operationalising.
  - Software/computing power.
- Level 4:
  - Raw data sources.

An Arthur-like conjecture might be that the innovations are likely to emanate from levels 3 and 4. In level 3 we have the opportunity to explore alternative hypotheses and to refine theories. Something like utility functions, profits and net benefits are likely to be present in some form or other to represent preferences with any maximisation hypotheses subject to a variety of constraints (which are themselves integral parts of theory-building). We might also conjecture that an underlying feature that is always present is that behaviour will be probabilistic. In fact this is likely to provide the means for integrating different approaches.

Can we identify likely innovation territories? The 'big and open data' movement will offer new sources, which will have impacts through levels 2 and 3. One consequence is likely to be the introduction of more detail – more categories – into the working model, exacerbating the 'number of variables' problem. This, in turn, could drive the modelling norm towards microsimulation. Such a development will be facilitated by increasing computing power. We are unlikely to have fundamentally different underlying hypotheses for theory-building, but there may well be opportunities to bring new mathematics to bear – particularly in relation to dynamics.

There is one other possibility of a different kind, reflected in level 2 system definition, in relation to scales. There is an argument that models at finer scales should be controlled by and made consistent with models at more aggregate scales. An obvious example is that the population distribution across the zones of a city should be consistent with aggregate level demography, and similarly for the urban economy. An intriguing possibility remains the application of the more aggregate methods (demographic and economic input–output) at fine zone scales.

Exercises

We have noted the power of disciplines and associated coalitions. We have seen how new disciplines (possibly subdisciplines) can evolve. Most importantly, however, we have seen how a system's focus forces interdisciplinarity in a fundamental sense. This leads to the notion of requisite knowledge in relation to our system of interest and our research challenges, followed by an exploration of combinatorial evolution. The last adds insight and leads us into constructing a map of the knowledge base for our research problem. With these foundations in mind, we can progress in the next chapter to some of the practicalities of interdisciplinary research, building on the framework presented in Chapter 1. The following exercises encourage exploration in your own research area.

1   If you were located in a particular discipline at some stage in your
     career, analyse its structure in terms of the STM and PDA frameworks
     (see Chapter 1). Select a research problem in this discipline and see
     if it can be usefully reformulated as an interdisciplinary problem.
     Are there any concepts from that discipline that could have roles as
     superconcepts?
2   Select a current research problem and apply the 'requisite knowledge
     test'. What do I need to know?
3   Take a system of interest and explore its hierarchical structure using
     the ideas of 'combinatorial evolution'.

Notes

1   Becher, *Academic Tribes and Territories*.
2   Wilson, *Knowledge Power*.
3   There have been attempts to change – Sussex, for example, in the UK – which largely failed, but
    somewhere like Arizona State University in the US 'abolished' departments. See R. Munck and
    K Mohrman (eds.) (2010) *Reinventing the university*, Glasnevin Publishing, Dublin referenced
    in Wilson, *Knowledge Power*.
4   In his Foreword to *Mathematics in the Social Sciences*, Stone makes these points very forcibly.
    This is especially important as it comes from an economist on the serviceableness of
    mathematics in the social sciences, he notes that 'the techniques developed for some specific
    purpose in one science can quite often be fruitfully applied in another'. Later he adds, 'My
    work on [the ... British economy] has brought home to me ... the difficulty of disengaging the
    economic aspects of life from their demographic, social and psychological setting.'
5   There are sometimes issues of 'esteem'. I recall an American economist observing to me that
    'putting the word "urban" before "economist" is rather like putting "horse" before "doctor"!'
6   See, for example, the illustrations in Wilson, *The Science of Cities and Regions*.
7   See Wilson, *Entropy in Regional and Urban Modelling*.
8   Lotka, *The Elements of Physical Biology*; Volterra, 'Population growth, equilibria and extinction
    under specific breeding conditions'; Richardson, *Arms and Insecurity*, which summarises his
    earlier work. See Chapter 4 of the present title.
9   Ashby, *An Introduction to Cybernetics*.
10 Weiner, *Invention*.
11 Arthur, *The Nature of Technology*.
12 Arthur, *The Nature of Technology*.
13 See Chapter 4, pages 38–40.

Part 2
Doing interdisciplinary research

Chapter 3
How to start

Introduction

A fruitful starting point is to apply the STM and PDA frameworks to a chosen system of interest. This process in itself will generate research problems and illustrations are provided below (pp. 27–30). It is suggested that it is often useful to begin with a simple description of the system and to build a 'toy' or 'demo' model of the system as the basis for the research. This is usually a feasible way of beginning the research process. It also provides a framework for adding depth as the research progresses (pp. 30–3). This is also a good stage to return to the issue of whether there are 'real challenges' (pp. 6–11) associated with your system and your proposed research, a question that is pursued here as 'research *on*' vs 'research *for*' (pp. 33–5).

First steps

There are starting points that we can take from systems thinking, theory development and methods – including data – that make up the STM framework (from Chapter 1). To recap:

- S: define the system of interest, dealing with the various dimensions of scale, etc.

- T: decide what kinds of theory, or understanding, can be brought to bear.
- M: determine what kinds of methods are available to operationalise the theory and to build a model.

This process is essentially analytical: how does the system of interest work? How has it evolved? What is its future? This approach will force an interdisciplinary perspective and within that force some choices. For example, statistics or mathematics? Econometrics or mathematical economics? We should also flag a connection to Brian Arthur's ideas in 2009 on the evolution of technology (see *Nature of Technology*) as applied to research (cf. pp. 20–22 above). He would argue that our system of interest in practice can be broken down into a hierarchy of subsystems, and that innovation is likely to come from lower levels in the hierarchy. This was, in his case, technological innovation, but it seems to me that this is applicable to research as well.

Then, if applicable, we have also seen a second step for getting started from Chapter 1: to ask questions about policy and planning in relation to the system of interest – the PDA. To recap once again:

- P: What is the policy (that is, what are the objectives for the future)? Should we develop a plan – another 'P'?
- D: Can we design (that is, 'invent') possible plans?
- A: We then have to test alternative plans by, say, running a model to analyse and evaluate them. Ideally the analysis would offer a range of indicators, perhaps using Sen's capability framework[1] or offering a full cost-benefit analysis.[2] A policy or a plan is, in formal terms, the specification of exogenous variables that can then be fed into a model-based analysis.[3]

These six steps form an important starting point that usually demands much thought and time. Note the links: the STM is essentially the means of analysis in the PDA. It may be thought that the research territory is in some sense pure analysis, but most urban systems of interest have real-world challenges associated with them and these are worth thinking about. Some ideas of research problems should emerge from this initial thinking. Some problems will arise from the challenges of model building, some from real, on-the-ground problems. See the following examples.

- Demographic models are usually built for an aggregate scale. Could they be developed for small zones – say for each of the 626 electoral wards in London?
- While there may be pretty good data on birth and death rates, migration proves much more difficult. First there are definitional problems: when is a move a migration – long distance? – and when is it residential relocation?
- If we want to build an input–output model for a city then, unlike the case at the national level, there will be no data on trade flows – imports and exports – so there is a research challenge to estimate these.
- There is then an economic analogue of the demography question: what would an input–output model for a small zone – say a London electoral ward – look like? This could be used to provide a topology of zone (neighbourhood) types – a new geodemographics.
- In the UK at the present time there is, in aggregate at the national level, a housing shortage. An STM description might focus on cities, or even on small zones within cities. How does the housing shortage manifest itself: through differential prices? What can be done about it? This last is a policy and planning question. Alternative policies and plans could be explored – the PD part of PDA – and then evaluated – the A part.
- What is the likely future of retail with regard to relative sizes of centres, the impact of internet shopping etc?
- How can 'parental choice' in relation to schools be made to work (without large numbers of people feeling very dissatisfied in obtaining only their second, third or fourth choices)?
- Can we – should we – aim to do anything about road congestion?
- Does responding to climate change at the urban scale involve shorter trips and higher densities? If so, how can this be brought about – the design-question? If not, why not?
- Can we speculate about the future of work in an informed way – taking account of the possibilities of 'hollowing out' through automation?

Research questions can be posed and the STM-PDA framework should help. The examples indicated are real and ambitious, and it is right that we should aim to be ambitious. However, given the resources that any of us have at our disposal, the research plan also has to be feasible. There are different ways of achieving feasibility, probably representing two poles of a spectrum: either narrow down the task to a small part of the bigger question or stay with the bigger question and try to break into it with something like a 'toy' model, to test ideas on a 'proof-of-concept' basis? The first strategy is the more conservative in approach and can be valuable. It is probably the most popular with undergraduates doing dissertations and postgraduate students – and indeed their supervisors. It is lower risk, but potentially less interesting.

We can then add a further set of basic principles – offering topics for thought and discussion once the STM-PDA analysis is done, at least in a preliminary way.

- Try to be comprehensive, to capture at least as much of the inevitable interdependence in your system of interest as is feasible.
- Review different approaches – e.g. to model building (which I tend to use) – and integrate these where possible. There are some good opportunities for spin-off research in this kind of territory.
- Think of applying good ideas more widely. I was well served in the use of the entropy concept in my early research days. Having started in transport modelling, because I always wanted to build a comprehensive model, I could apply the concept to other subsystems and (with Martyn Senior) find a way of making an economic residential location model optimally sub-optimal.[4]
- The 'more widely' also applies to other disciplines. Modelling techniques that work in a contemporary situation, for example, can be applied to historical periods – even ancient history and archaeology. (See Chapter 7, pp. 89–91).
- There is usually much work to do on linking data from different sources and making it fit the definitions of your system of interest. Models can also be used for estimating missing data and for making samples comprehensive.

Preliminary thinking is done and some structure generated. Now it is time to get started.

Adding depth

A valuable principle for starting a piece of research is to start simple and then, progressively, add depth. We illustrate this with an urban example.

An appropriate ambition of the model-building component of urban science is the construction of the best possible comprehensive model which represents the interdependencies that make cities complex (and interesting) systems. To articulate this is to spell out a kind of research programme – how do we combine the best of what we know into such a general model? Most of the available 'depth' is in the application of particular submodels, notably transport and retail. If we seek to identify the 'best' – and there are many subjective decisions to be made here in a contested area – we define a large-scale computing operation underpinned by a substantial information system that houses relevant data. Though a large task, this is feasible. How would we set about it?

The initial thinking through would be an iterative process. The first step would be to review all the submodels, especially their categorisation of their main variables – their system definitions. Almost certainly these would not be consistent: each would have detail appropriate to that system. It would then be necessary – or would it? – to find a common set of definitions. It may be possible to work with different classifications for different submodels, then to integrate them in some way, thus connecting the submodels as part of a general model that, among other things, captures the main interdependencies. This is a research question in itself. It is at this point that it would be necessary to confront the question of exogenous and endogenous variables. We want not only to maximise the number of endogenous variables, but also to retain as exogenous those that will be determined externally, for example by a planning process.

There is then the question of scales and possible relations between scales. Suppose we can define our city, or a city region, divided into zones, with an appropriate external zone system (including a 'rest of the world' zone to close the system). Then, for the city in aggregate, we would normally have a demographic model and an economic model. These would provide controlling totals for the zonal models: for example, the zonal populations would add up to those of the aggregate demographic model. There is also the complicated question of whether we would have two or more zone systems – say one with larger zones, one with finer-scale zones. For simplicity at this stage, however, assume we are using a one zone system. We can then begin to review the submodels.[5]

The classic transport model has four submodels: trip generation, distribution, modal split and assignment. As this implies, the model includes a multi-modal network representation. Trips from origin to destination by mode (and purpose) are loaded onto the network. This enables congestion to be accounted for and properly represented in generalised costs (with travel time as an element) – a level of detail not normally captured in the usual running of spatial interaction models.

A fine-grain retail model functions with a detailed categorisation of consumers and of store attractiveness; it can then predict flows into stores with reasonable accuracy. This model can be applied in principle to any consumer-driven service – particularly flows into medical facilities, for example, and especially general practice surgeries. This task is different if the flows are assigned by a central authority as to schools, for instance.

The location of economic activity, and particularly employment, is more difficult. Totals by sector might be derived from an input–output model, but the numbers of firms are too small to use statistical averaging techniques. What ought to be possible with models is to estimate the relative desirability of different locations for different sectors, then to use this information to interpret the marginal location decisions of firms. This approach fits with the argument about the full application of urban science being historical, which is discussed below.

In all cases of location of activities, it will be necessary in a model to incorporate constraints at the zonal scale, particularly in relation to land use. As these are applied, measures of 'pressure', for example on housing at particular locations, can be calculated (and related to house prices). It is these measures of pressure that lie at the heart of dynamic modelling, and it is to this that we will turn shortly.

The Lowry model, which we will describe in more detail in Chapter 4 (pp. 38–43) was comprehensive, built on only 12 equations and oversimplified – but it showed how to 'break in' to a complex research challenge. As this sketch indicates, it would be possible to construct a Lowry-like model that incorporated the best-practice level of detail from any of the submodels. Indeed, it is likely that within the hardy band of comprehensive modellers – Marcial Echenique, Michael Wegener, Roger Mackett, Mike Batty and David Simmonds, for example – this will largely have been done, though my memory is that this is usually without a full transport model as a component. What has not been done, typically, is to make these models fully dynamic. Rather, in forecasting mode, they are run as a series of equilibrium positions, usually on the basis of changes that are exogenous to the model.[6]

The next step is to build a comprehensive model, incorporating key interdependencies, that is fully dynamic. This has been attempted by Joel Dearden and myself as reported in Chapter 4 of *Explorations in Urban and Regional Dynamics*.[7] However, it should be emphasised that this is a proof-of-concept exploration and does not contain the detail – for instance on transport – that is being advocated above. It does tackle the difficult issues of moves: non-movers, job movers, house movers, house and job movers, which is an important level of detail in a dynamic model but very difficult to handle in practice. It also attempts to handle health and education explicitly, in addition to conventional retail. Significantly, as a nonlinear model, it does embrace the possibility of path-dependent phase changes. These are illustrated a) by the changes in initial conditions that would be necessary to revive a High Street and b) in terms of gentrification in housing.

What can we learn from this sketch? First, it is possible to add much more detail than is customary, but this is difficult in practice. I would conjecture this is because to do so effectively demands a substantial team and corresponding resources; in contrast to particle physics, these kinds of resources are not available to urban science. Second, and rather startlingly, it can be argued that the major advance of this kind of science will lie in urban *history*. This is because in principle all the data is available, even that which we have to declare exogenous from a modelling perspective. The exogenous variables can be fed into the model and the historians, geographers and economic historians can interpret their evolution. This would demand serious team work, but would be the equivalent for urban science of unravelling DNA in biology or demonstrating the existence of the Higgs boson in particle physics. Where are the resources – and the ambition – for this?

'Research on' vs 'research for'

The previous section adds substance to the framework discussed in Chapter 1 (pp. 4–6). In a similar way we can explore a new perspective on the 'real challenges' as introduced there (pp. 6–11). In the usual way we can begin with a system of interest, henceforth 'the system' for brevity. We can then make a distinction between the science of the system and the applied science relating to the system. In the second case, the implication is that the system offers challenges and problems with which the science (of that system, or possibly also with associated systems) might assist. In research terms, this distinction can be roughly classified as 'research *on*' the system and 'research *for*' the system. This might be physics on the one hand and engineering on the other; or biological sciences on the one hand and medicine on the other. There are many groups of disciplines like this where there is a division of labour, though whether this division is always either clear or efficient can be a matter of debate. In the case of urban research (and possibly the social sciences more generally), perhaps because it is an under-developed interdisciplinary area, there is a division of labour, but one that leaves a significant grey area. In the case of cities, the practitioners (the planners) are not well served by the 'research on' community – or perhaps they are not sufficiently well equipped to engage.[8] But there is also a concern that the division is too sharp and that the balance of research effort is focused more strongly on 'research on' rather than contributing to 'research for'.

There are a number of complications that we have to work to resolve. First, there is the fact that there are disciplinary agendas on cities, for example in economics, geography and sociology, where they ought to be interdisciplinary. However, this does illustrate the fact that there is a 'research on' versus 'research for' challenge. The 'research on' school are concerned with how cities work, the 'research for' group with, for example, how to 'solve' (if that is the right question) issues of traffic congestion or housing problems or social disparities. As we have already seen, it is a long list.

A second complicating issue in the UK is the research councils' 'impact' agenda. I have no problem with a requirement that all research should be intended to have impact – the opposite is absurd. However, that depends on the possibility of the impact being intellectual impact within the science; that is, impact within 'research on'. What seems to have happened is that the research councils' definition has narrowed and impact in their sense is intended to relate to 'real' problems – in other words, to 'research for'. Consider physics and engineering: while the toolkits overlap in some respects, they, and the associated mindsets, are pretty different. The same could be argued for research on cities except in this case, we do not have labels that are analogous to physics and engineering. We therefore have to invent our own. From a research council perspective, this has not been clearly handled. There is an expectation that for any application, there will be a 'pathways to impact' statement. If the research in question is of a 'research on' kind, and if the associated tools do not obviously fit 'research for', then this is very difficult and quite a lot of jumping through hoops is required.

A third issue is the influence of the Research Excellence Framework (REF) on research priorities in the UK.[9] Again there is an element of required impact and yet the bulk of the panels are made up of 'research on' academics. It is even argued – or is it just in our subconscious? – that 'pure' research is more worthwhile in REF terms than 'applied'. It was once suggested to me in the context of a university business school – although not in these words – that the 'research on firms' was more important for the REF than 'research for firms', because the latter could be considered as consultancy and therefore of a lower grade. There is some truth in this in that 'research on' can produce wider ranging, general results that offer insight, as opposed to specific case studies that do not generalise. However, in the social sciences at least, it is the case studies that eventually lead to the general, grounded in evidence.

There is then a fourth issue, more like a challenge: if impact is really desirable (and it is) how can the users get the best from the researchers? It is often argued that the UK has very high-quality research but that it fails, to a substantial extent, to reap the rewards through application. Indeed, there have been commissions of many kinds for decades on how academic research can be better linked to application; I would guess a study roughly once every two years. There are various 'solutions' and many have been tried, but success has been partial at best. The 'research on' community remains the largest group of academic researchers and retains the 'prestige' that serves it well in many ways. There are significant straws in the wind, at least in the UK: a shifting of research resources in the direction of Innovate UK and the establishment of the catapults[10], and some redirection of research council funding. Yet my guess is that there is a battle for hearts and minds that is still being fought.

What do we need? Some clarity of thought, some changes of mindset, especially in terms of prestige, and – perhaps above all – some demonstrators that show that 'research for' can be just as exciting as 'research on' – in many cases much more so. In the urban research world, we are in principle fortunate in that we can have it both ways. Discoveries in the science often have pretty immediate applications, but there are opportunities for more 'research on' researchers to spend at least some time in the 'research for' community. In my own case, the most striking example was working to build a spin-out company, GMAP, as presented in Chapter 6 (pp. 75–77). This was a demonstration of having it both ways – 'research for' provided access to data which could then be used in basic research ('research on').

Exercises

We are now in a position to start. We can define a research challenge and we have frameworks that help us to be interdisciplinary.

1 Apply the STM and PDA frameworks to a system of interest of your choice and begin to define a possible research project.
2 Think of the simplest way of setting up your problem for a quick start and plan for adding depth later.
3 Are there 'real challenges' associated with your system of interest and proposed research? Should you make this central to your research or think of it as a bonus?

Notes

1   For example, see Sen, *Commodities and Capabilities*.
2   See Foster and Beesley, 'Establishing the social benefit of constructing an underground railway in London'.
3   See Wilson, *The Science of Cities and Regions* for an overview; Wilson, 'New roles for urban models'.
4   Wilson and Senior, 'Some relationships between entropy maximising models, mathematical programming models and their duals'.
5   Roumpani, unpublished PhD thesis, UCL.
6   See Batty and Milton, 'A new framework for very large-scale urban modelling'.
7   Dearden and Wilson, *Explorations in Urban and Regional Dymanics*.
8   Recall Britton Harris's comment on PDA that there are these three kinds of thinking and you rarely find all three in the room at the same time.
9   The 'Research Excellence Framework' (REF) which has been in place in the UK in some form or other since 1987.
10 'Catapults' are technology transfer organisations established by Innovate UK.

Chapter 4
Establishing a research base 1: system models

Introduction: models, then data

In broad terms, 'analysis' represents the underpinning science via STM, applied through the PDA framework. In the illustrations that follow, the emphasis is on quantitative research to illustrate the principles of interdisciplinary research, fully recognising that there are alternative, relevant and complementary perspectives and that indeed the qualitative informs the quantitative, particularly through the theory-building part of STM. However, the development of computing power and, more recently, access to real-time 'big data' sources, have added momentum to quantitative interdisciplinary social, economic and geographic research. A narrative on a theory of how a city functions, for example, can be translated into mathematics and then into a computer model. In a number of sections below we use the idea of a model to illustrate interdisciplinary urban research that can also be applied within a PDA framework.

As an example, the history of urban mathematical and computer modelling is outlined (pp. 38–43) as a prime example of interdisciplinary research. Of course, there are different and competing approaches to model development. These are explored in ways that might help researchers make their choices in model design (pp. 43–47 and 47–49). If the research includes building mathematical models of the system of interest, it is interesting and potentially useful to explore equations that have been developed for particular contexts but that turn out to have applications across disciplines. Examples are offered as 'equations with names' (pp. 49–52).

We will continue this discussion in Chapter 5 where we explore data sources for research. However, the order of these chapters is important. System definitions and the articulation of the research problem should come first – models, then data.

The power of modelling: understanding and planning cities

The 'science of cities' has a long history. The city was the market for von Thunen's analysis of agriculture in the late eighteenth century while many largely qualitative explorations took place in the first half of the twentieth century. However, cities are complex systems and the major opportunities for scientific development came with the emergence of computer power in the 1950s. This coincided with large investments in highways in the United States. Computer models were developed to predict both current transport patterns and the impacts of new highways and land use plans. Plans could be evaluated and the formal methods of what became cost-benefit analysis were developed around that time. However, it was always recognised that transport and land use were interdependent and that a more comprehensive model was needed. Several attempts were made to build such models, but the example that stands out is Ira S. Lowry's *Model of Metropolis*, published in 1964. This model was elegant and (deceptively) simple, represented in just 12 algebraic equations and inequalities. Many contemporary models are richer in detail and have many more equations, but most are Lowry-like in that they have a recognisably similar core structure.[1]

So what is this core and what do we learn from it? How can we enhance our understanding by adding detail and depth? What can we learn by applying contemporary knowledge of system dynamics? What does all this mean for future policy development and planning? The argument is illustrated and referenced from my own experience as a mathematical and computer urban modeller, but the insights work on a broader canvas.

The Lowry model is iconic in its representation of urban theory into a comprehensive model. He[2] started with some definitions of urban economies with two broad categories: 'basic' – mainly industry and, from the city's perspective, exporting; and 'retail', broadly defined to mean anything that serves the population.[3] Lowry then introduced some key hypotheses about land. For each zone of the city he took total land, identified unusable land, allocated land to the basic sector and then argued that the rest was available to retail and housing, with retail having priority. Land available for housing, therefore, was essentially a residual.

A model run then proceeds iteratively. Basic employment is allocated exogenously to each zone – possibly as part of a plan. This employment is then allocated to residences and converted into total population in each zone. This link between employment zones and residential zones can be characterised as 'spatial interaction' manifested by the 'journey to work'. The population then 'demands' retail services and this generates further employment, which is in turn allocated to residential zones. (This is another spatial interaction – between residential and retail zones.) At each stage in the iteration the land use constraints are checked. If they are exceeded (in housing demand), the excess is reallocated. And so the city 'grows'. This growth can be interpreted as the model evolving to an equilibrium at a point in time or as the city evolving through time – an elementary form of dynamics.

The essential characteristics of the Lowry model which remain at the core of our understanding are:

- The distinction between basic (outward-serving) and retail (population-serving) sectors of the urban economy.
- The 'spatial interaction' relationships between work and home and between home and retail.
- The demand for land from different sources, and in particular housing, being forced to greater distances from work and retail as the city grows. This has obvious implications for land value and rents.

In the half century since Lowry's work was published, depth and detail have been added and the models have become realistic, at least for representing a working city and for short-run forecasting. The longer run still provides challenges, as we will see. It is now more likely that the Lowry model iteration would start with some 'initial conditions' that represent the current state. The model would then represent the workings of the city and could be used to test the impact of investment and planning policies in the short run. The economic model and the spatial interaction models would be much richer in detail and, while it remains non-trivial to handle land constraints, submodels of land value both help to handle this and are valuable in themselves.[4]

Specifically:

- The key variables can all be disaggregated – people, for example, can be characterised by age, sex, education attainment and skills. They may thus be better matched to a similarly disaggregated set of economic sectors, demanding a variety of skills and offering a range of incomes.
- Population analysis and forecasting can be connected to a fully-developed demographic model.
- The economy can be described by full input–output accounts and the distinction between basic and retail can be blurred through disaggregation.
- The residential location component can be enriched through the incorporation of utility functions with a range of components. House prices can be estimated through estimates of 'pressure', thus facilitating the effective modelling of what types of people live where.
- All this serves to reinforce the idea that the different elements of the city are interdependent.

'Housing pressure' will be related to the handling of land constraints in the model. In the Lowry case this was achieved by the reallocation of an undifferentiated population when zones became 'full'. With contemporary models, because house prices can be estimated (or some equivalent), it is these prices that handle the constraints.

While the Lowry-type models remain comprehensive in their ambition, sectoral models – particularly in the transport and retail cases – are usually developed separately in even greater depth; as such, they can be used for short-run forecasting. Supermarket companies, for example, routinely use such models to estimate the revenue attracted to proposed new stores, which supports the planning of their investment strategies.[5]

The models as described above are essentially statistical averaging models.[6] They work well for large populations where the predictions of the models are of 'trip bundles' rather than of individual behaviour. The models work well precisely because of this averaging process, which removest the idiosyncrasies of individuals. They use the mathematics developed by Ludwig Boltzmann for physics in the late nineteenth century, but with a different theoretical base. What can we then say about individual behaviour, however? Two things: we can interpret the 'averaging models' and we can seek to model individual behaviour directly, albeit probabilistically.

In the first case, elements of the models can be interpreted as individual utility functions. In the retail case, for example, it is common to estimate the perceived benefits of shopping centre size and to set these against the costs of access (including money costs and estimated values of different kinds of time). What the models do through their averaging mechanism is to represent distributions of behaviour around average utilities. This is much more realistic than the classic economic utility maximising models, as shown through goodness-of-fit measures. In effect, the classic model in the retail case would assume that everyone travels to their nearest centre. Empirically it is clear that this is not the case, whether through imperfect information for the consumer or simply a range of preferences that have not been captured in the classic model.

The second case demands a new kind of model, and these have been developed as so-called agent-based models (ABMs). A population of individual 'agents' is constructed along with an 'environment'. The agents are then given rules of behaviour and the system evolves. If the rules are based on utility maximisation on a probabilistic basis, then the two kinds of model can be shown to be broadly equivalent.[7]

The argument to date has been essentially geo-economic, though with some implicit sociology in the categorisation of variables when the models are disaggregated. There is more depth to be added in principle from sociological and ethnographic studies; if new findings can be clearly articulated, this kind of integration can be achieved.

The models described thus far represent the workings of a city at a point in time – give or take the dynamic interpretation of the Lowry model. There is an implicit assumption that if there is a 'disturbance' – an investment in a new road or a housing estate, for example – then the city returns to equilibrium very quickly; this can thus be said to characterise the 'fast dynamics'. It does mean that these models can be used – and indeed are used – to estimate the impacts of major changes in the short term. The harder challenge is the 'slow dynamics', which seek to model the evolution of the slower changing structural features of a city over a longer period. This takes us into interdisciplinary territory, sometimes named as 'complexity science'. When the task of building a fully dynamic model is analysed, it becomes clear that there are nonlinear relationships – as retail centres grow, for example, there is evidence that there are positive returns to scale. Technically we can then draw on the mathematics of nonlinear complex systems. These show that we can expect to find path dependence – that is, dependence on initial conditions – and phase changes – that is, abrupt changes in form as 'parameters' (features such as income or car ownership) – pass through critical values.

The particular models in mathematical terms bear a family relationship to the Lotka-Volterra models.[8] These were originally designed to model ecological systems in the 1930s, but can now be seen as having a much wider range of application (see also pp. 49–52).[9]

These ideas can be illustrated in terms of retail development. In the late 1950s and early 1960s, corner-shop food retailing was largely replaced by supermarkets. By the standards of urban structural change, this was very rapid. It can be shown that this situation arose though a combination of increasing incomes and car ownership, which in effect increased the accessibility of more distant places. This was a phase change. Path dependence is illustrated by the fact that if a new retail centre is developed, its success in terms of revenue attracted will be dependent on the existing pattern of centres – the initial conditions. Again this can be analysed using dynamic models.[10]

This leads us to two fundamental insights. First, it is impossible to forecast for the long term because of the likelihood of phase changes at some point in the future. Second, the initial structure of the city – the 'initial conditions' – might be thought of as the 'DNA' of the city; this will in substantial part determine what futures are possible. Attempts to plan new and possibly more desirable futures can be thought of as 'genetic planning' by analogy with genetic medicine.

Given these insights, how can we investigate the long term – 25 or 50 years into the future? We can investigate a range of futures through the development of scenarios and then we can deploy Lowry-Boltzmann-like models to investigate the performance of these. We can also use the fully dynamic Lotka-Volterra models to explore the possible paths to give insights on what has to be done to achieve these futures.

There is a key distinction in the application of models: one between the variables that are *exogenous* to the model and those that are *endogenous*. The exogenous variables are specified either as forecasts or as components of plans, and the model can then be run to calculate the endogenous variables for the new situation. This is done more or less routinely in transport and retail sector planning. For example, a new road can be 'inserted' into the model, the model rerun and the 'adjusted' city explored. In this case, a cost-benefit analysis can be carried out, along with the calculation of accessibilities. In the case of retail, a developer or a retailer can run the model to calculate the revenue attracted to a new store, then calculate the maximum level of investment that would make such a store profitable – often to determine how much to bid for a site. It is possible now, but relatively rare, to apply these methods in the public sector in fields such as education and health. Indeed, model-based methods could be used to underpin master planning and thus contribute to effective housing development and associated green belt policies.

As we have seen in our brief review of dynamics, this only works in this way for the short run – the impacts of building a new road or opening a new store. For the long run, it is necessary to shift to scenario development. In doing so we create opportunities to explore possible solutions to the biggest challenges – the so-called 'wicked problems'.

It is not difficult to construct a list of wicked problems.[11] In the UK, for example, this might include regeneration, especially economic, in many 'poor' towns; embracing the north vs south issues and opportunities for fulfilling work more broadly; chronic housing shortages; transport congestion, limiting accessibilities; the long tail of failure in education; the postcode lottery aspect of health care; and, perhaps the biggest challenge of all, responding to climate change and low-carbon targets. Most of this list and more, with appropriate variations, will be evident in most countries. Applications of computer models will not solve these problems. This brings home the policy and design dimensions of planning: policies to attack wicked problems need to be 'serious' and to be seen to be so; possible solutions have to be invented. These ambitions then provide the basis for developing more radical scenarios as well as 'more of the same'. And then the skills of the modeller kick in again in the analysis of feasibility, calculating costs and benefits, and charting the path from A to B – from the present to a rewarding future.

Competing models: truth is what we agree about

I have always been interested in philosophy and the big problems – particularly 'What is truth?'. How can we know whether something – a sentence, a theory, a mathematical formula – is true? And I guess because I was a mathematician and a physicist early in my career, I was particularly interested in the subset of this which is the philosophy of mathematics and the philosophy of science. I read a lot of Bertrand Russell, which perhaps seems rather quaint now.[12] The maths and the science took me into Popper and the broader reaches of logical positivism. Time passed and I found myself a young university professor working on mathematical models of cities, then the height of fashion. Fashions change, however, and by the late 1970s, on the back of distinguished works such as David Harvey's *Social Justice and the City*, I found myself under sustained attack from a broadly Marxist front. 'Positivism' became a term of abuse and Marxism, in philosophical terms – or at least my then understanding of it – merged into the wider realms of structuralism. I began to understand that there were hidden (often power) structures to be revealed in social research that the models I was working on missed, therefore weakening the results.

This was serious stuff. I could reject some of the attacks in a straightforward way. There was a time when it was argued that anything mathematical was positivist and therefore bad and/or wrong. This argument could be rejected on the grounds that mathematics was a tool and that there were indeed distinguished Marxist mathematical economists such as Piero Sraffa. But I had to dig deeper in order to understand. I read Marx and a lot of structuralists, some of whom, at the time, were taking over English departments. I even gave a seminar in the Leeds English Department on structuralism.

In my reading, I stumbled on Jurgen Habermas which was a revelation for me. His work took me back to questions about truth and provided a new way of answering them. In what follows, I am sure I oversimplify. His work is very rich in ideas, but I took a simple idea from it: truth is what we agree about. I say this to students now who are usually pretty shocked. But let us unpick it. We can agree that $2 + 2 = 4$. We can agree about the laws of physics – up to a point anyway: there are discoveries to be made that will refine these laws, as has happened in the past. That also connects to another idea that I found useful in my toolkit: C. S. Peirce and the pragmatists. I will settle for the colloquial use of 'pragmatism': we can agree in a pragmatic sense that physics is true and handle the refinements later. I would argue from my own experience that some social science is 'true' in the same way: much demography is true up to quite small errors – think of what actuaries do. But when we get to politics, we disagree. We are in a different ballpark. We can still explore and seek to analyse, and having the Habermas distinction in place helps us to understand arguments.

How does the agreement come about? The technical term used by Habermas is 'intersubjective communication' and there has to be enough of it. In other words, the 'agreement' comes on the back of much discussion, debate and experiment. This fits very well with how science works. A sign of disagreement is when we hear that someone has an 'opinion' about an issue. This should be the signal for further exploration, discussion and debate rather than simply a 'tennis match' kind of argument.

Where does this leave us as social scientists? We are unlikely to have laws in the same way that physicists have laws, but we do have truths, even if they are temporary and approximate. We should recognise that research is a constant exploration in a context of mutual tolerance – our version of intersubjective communication. We should be suspicious of the newspaper article which begins 'research shows that' when the research quoted is based on a single, usually small, sample survey and regression analysis. We have to tread a line between offering knowledge and truth on the one hand and recognising the uncertainty of our offerings on the other. This is not easy in an environment where policy makers want to know what the evidence is, or what the 'solution' is, for pressing problems and would like to be more assertive than we might feel comfortable with. The nuances of language to be deployed in our reporting of research become critical.

Models are representations of theories. I write this as a modeller – someone who works on mathematical and computer models of cities and regions, but who is also seriously interested in the underlying theories I am trying to represent. My field, relative say to physics, is underdeveloped. This means that we have a number of competing models and it is interesting to explore the basis of this and how to respond. What is 'truth' in this context? There may be implications for other fields – even physics.

A starting conjecture is that there are two classes of competing models: a) those that represent different underlying theories (or hypotheses); and b) those that stem from the modellers choosing different ways of making approximations in seeking to represent very complex systems. The two categories overlap, of course. I will conjecture at the outset that most of the differences lie in the second (with perhaps one notable exception). So let us get the first out of the way. Economists want individuals to maximise utility and firms to maximise profits – simplifying somewhat, of course. They can probably find something that public services can maximise – health outcomes, exam results – indeed a whole range of performance indicators. There is now a recognition that for all sorts of reasons, the agents do not behave perfectly and ways have been found to handle this. There is a whole host of (usually) micro-scale economic and social theory that is inadequately incorporated into models, in some cases because of the complexity issue – the finer realities are approximated away; but in principle that can be handled and should be. There is a broader principle lurking here: for most modelling purposes, the underlying theory can be seen as maximising or minimising something. So if we are uncomfortable with utility functions, or with economics more broadly, we can still try to represent behaviour in these terms – if only to have a base line from which behaviour deviates.

So what is the exception – another kind of dividing line which should perhaps have been a third category? At the pure end of a spectrum, 'letting the data speak for themselves'. It is mathematics vs statistics; or econometrics vs mathematical economics. Statistical models look very different – at least at first sight – to mathematical models and usually demand quite stringent conditions to be in place for their legitimate application. Perhaps, in the quantification of a field of study, statistical modelling comes first, followed by the mathematical? There is of course a limit in which both 'pictures' can merge: many mathematical models, including the ones I work with, can be presented as maximum likelihood models.

There is perhaps a second high-level issue. It is sometimes argued that there are two kinds of mathematician: those who think in terms of algebra and those who think in terms of geometry. (I am in the algebra category, which I am sure biases my approach.) As with many of these dichotomies, they should be removed and both perspectives fully integrated. But this is easier said than done.

How do the 'approximations' come about? I once tried to estimate the number of variables I would like to have for a comprehensive model of a city of one million people; at a relatively coarse grain the answer was around 10^{13}.[13] This demonstrates the need for approximation in the system representation. The first steps can be categorised in terms of scale: first, spatial – referenced by zones of location rather than continuous space – and how large should the zones be? Second, temporal: continuous time or discrete? Third, sectoral: how many characteristics of individuals or organisations should be identified and at how fine a grain? Experience suggests that the use of discrete zones – and indeed other discrete definitions – makes the mathematics much easier to handle. Economists often use continuous space in their models, for example, and this forces them into another kind of approximation, monocentricity, which is hopelessly unrealistic. Many different models are simply based on different decisions about, and representations of, scale.

The second set of differences turn on focus of interest. One way of approximating is to consider a subsystem such as transport and the journey to work, or retail and the flow of revenues into an individual shop or a shopping centre. The dangers here are that the critical interdependencies are lost and this always has to be borne in mind. Consider the evaluation of new transport infrastructure, for example. If this is based purely on a transport model, there is a danger than the cost-benefit analysis will be concentrated on time savings rather than the wider benefits – perhaps represented by accessibilities. There is also a potentially higher-level view of focus. Lowry once very perceptively pointed out that models often focus on activities – and the distribution of activities across zones; or on the zones, in which case the focus would be on land use mix in a particular area. The trick, of course, is to capture both perspectives simultaneously, something that Lowry achieved himself very elegantly, but that has been achieved only rarely since.

A major bifurcation in model design turns on the time dimension and the related assumptions about dynamics. Models are much easier to handle if it is possible to make an assumption that the system being modelled is either in equilibrium or will return to a state of equilibrium quickly after a disturbance. There are many situations where the equilibrium assumption is pretty reasonable – for representing a cross-section in time or for short-run forecasting, for example, representing the way in which a transport system returns to equilibrium after a new network link or mode is introduced. But, as we have seen, the big challenge is in the 'slow dynamics': modelling how cities evolve.

It is beyond our scope here to review a wide range of examples. If there is a general lesson, it is that we should be tolerant of one another's models; we should be prepared to deconstruct them to facilitate comparison and perhaps to remove what appears to be competition but need not be. The deconstructed elements can then be seen as building bricks that can be assembled in a variety of ways. For example, 'generalised cost' in an entropy-maximising spatial interaction model can easily be interpreted as a utility function and therefore not in competition with economic models. Cellular automata models and agent-based models are similarly based on different 'pictures' – different ways of making approximations. There are usually different strengths and weaknesses in the different alternatives. In many cases, with some effort, they can be integrated.[14] From a mathematical point of view, deconstruction can offer new insights. We have, in effect, argued that model design involves making a series of decisions about scale, focus, theory, method and so on. What will emerge from this kind of thinking is that different kinds of representations – 'pictures' – have different sets of mathematical tools available for the model building. Some of these are easier to use than others and so, when this is made explicit, might guide the research planning process.

Model components: abstract modes

I spent much time working on the Government Office for Science Foresight project, The Future of Cities. The focus was on the basis of a time horizon of 50 years into the future. It is clearly impossible to use urban models to forecast such long-term futures, but it is possible in principle systematically to explore a variety of future scenarios. A key element of such scenarios is transport; we have to assume that what is on offer – in terms of modes of travel – will be very different to today, not least to meet sustainability criteria. The present dominance of car travel in many cities is likely to disappear. How, then, can we characterise possible future transport modes?[15]

This takes me back to ideas that emerged in papers published 50 years ago (or, in one case, almost that). In 1966 Dick Quandt and William Baumol,[16] distinguished Princeton economists, published a paper in the *Journal of Regional Science* on 'abstract transport modes'. Their argument was precisely that in the future technological change would produce new modes: how could they be modelled? Their answer was to say that models should be calibrated not with modal parameters, but with parameters that related to the *characteristics* of modes. The calibrated results could then be used to model the take up of new modes that had new characteristics. By coincidence Kelvin Lancaster, an economist from Columbia University, published a paper, also in 1966, in *The Journal of Political Economy* on 'A new approach to consumer theory'[17] in which he defines utility functions in terms of the characteristics of goods rather than the goods themselves. Lancaster elaborated this in 1971 in his book *Consumer demand: A new approach*. In 1967 my 'entropy' paper ('A statistical theory of spatial distribution models') was published in the journal *Transportation Research* and a concept used in this was that of 'generalised cost'. This assumed that the cost of travelling by a mode was not just a money cost, but the weighted sum of different elements of (dis)utility: different kinds of time, comfort and so on, as well as money costs. The weights could be estimated as part of model calibration. In their magisterial history of transport modelling *Forecasting Urban Travel*, David Boyce and Huw Williams wrote, quoting my 1967 paper, that

> impedance ... may be measured as actual distance, as travel time, as cost, or more effectively as some weighted combination of such factors sometimes referred to as generalised cost. ... In later publications, 'impedance' fell out of use in favour of 'generalised cost'.[18]

(They kindly attributed the introduction of 'generalised cost' to me.)

This all starts to come together. The Quandt and Baumol 'abstract mode' idea has always been in my mind and I was attracted to the Kelvin Lancaster argument for the same reasons – though that does not seem to have taken off in a big way in economics. (I still have his 1971 book, purchased from Austicks in Leeds for £4.25.) At the time I never quite connected 'generalised cost' to 'abstract modes', but I certainly do now.

When we have to look ahead to long-term future scenarios, it is potentially valuable to envisage new transport modes in generalised cost terms. By comparing one new mode with another, we can make an attempt – albeit an approximate one, because we are transporting current calibrated weights forward 50 years – to estimate the take up of modes by comparing generalised costs. I have not yet seen any systematic attempt to explore scenarios in this way and I think there is some straightforward work to be done – do-able in an undergraduate or a master's thesis.

We can also look at the broader questions of scenario development. Suppose, for example, we want to explore the consequences of high-density development around public transport hubs. These kinds of policies can be represented in our comprehensive models by constraints – and I argue that the idea of representing policies – or more broadly 'knowledge' – in constraints within models is another powerful tool. This also has its origins in a 50-year-old paper – Ira S. Lowry's *Model of Metropolis*,[19] as introduced previously (pp. 38–40). In broad terms, this represents the fixing through plans of a model's exogenous variables. However, the idea of 'constraints' implies that there are circumstances where we might want to fix what we usually take as endogenous variables.

So we have the machinery for testing and evaluating long-term scenarios – albeit building on 50-year-old ideas. It needs a combination of imagination – thinking what the future might look like – and analytical capabilities – 'big modelling'. It is all still to play for, but there are some interesting papers waiting to be written.

Equations with names: the importance of Lotka and Volterra (and Tolstoy?)

One approach to developing ideas to support interdisciplinarity is to look at those 'equations with names' that over time have established themselves beyond their founders and initial purposes into a more public (at least in academia) consciousness. Indeed, it can be argued that there is a Darwinian process that allows these to emerge as being both game changers and having wider roles. In this section I draw on my own experience as an indicator of where we can continue to pick up new ideas from these kinds of equations.

The most famous equations with names – at least one being known almost universally – seem to come from physics. Newton's Law of Gravity – the gravitational force between two objects is proportional to their masses and inversely proportional to the distance between them; Maxwell's equations for electromagnetic fields; the Navier-Stokes's equation in fluid dynamics; and $E = mc^2$, Einstein's equation which converts mass into energy. The latter is the only equation to appear in the index (under 'E') in Ian Stewart's book *Seventeen equations that changed the world*.[20] While the gravitational law has been used to represent situations where distance attenuation is important, the translation is analogous and not exact. An interesting example, pointed out to me by Mark Birkin, is that of Tolstoy in *War and Peace*, written in the 1860s:

> Meanwhile, the very next morning after the battle, the French army moved against the Russians, carried along by its own impetus, now accelerating in inverse proportion to the square of the distance from its goal.[21]

The physics equations, on the whole, work in physics and not elsewhere. An exception – that is, one that does work elsewhere and has served me well in my own work – is Boltzmann's equation for entropy, $S = k \log W$ (to be found on his gravestone in Vienna and on many book covers, including one of mine). The other equations which have served me well – plural because they come in several forms – are the Lotka-Volterra equations, originally developed in ecology. Because of the nature of ecology relative to physics, they do not deliver the physics kind of 'exactness'. However, this may in part be the reason for their utility in translation to other disciplines.

The Boltzmann entropy-maximising method[22] works for any problem (and hence in a variety of fields) where there are large numbers of weakly interacting elements and where interesting questions can be posed about average properties of the system. Boltzmann does this for the distribution of energy levels of particles in gases at particular temperatures, for example. In my own work I use the method to calculate, for instance, journey to work flows in cities. The entropy measure was also introduced by Shannon into information theory. In one sense it underpins much of computer science through the notion of the 'bit', and is the basis of much of the development of electronic communications engineering. When Shannon produced his equation to measure 'information', he is said to have consulted the famous mathematician John von Neumann on what to call the main term. 'Call it "entropy",' von Neumann replied (paraphrased). He went on:

> It is like the entropy in physics, and if you do this you will find in any argument no one will understand it and you will always win.[23]

I would dare to say that von Neumann was wrong in this respect: it can be explained. In his recent book *Why Information Grows*,[24] Cesar Hidalgo makes an interesting point about Boltzmann's work – it crosses and links scales with the atoms in the micro with the thermodynamic properties of the macro. This was unusual at the time and perhaps still is.[25]

The Lotka-Volterra equations are concerned with systems of populations of different kinds – different species in ecology, for example. In one sense their historical roots can be related back to Thomas Malthus and his exponential 'growth of population' model. In that model there were no limits to growth; these were added by the Belgian mathematician Pierre François Verhulst, who dampened growth to produce the well-known logistic curve. In the 1970s Bob May showed that this simple model has remarkable properties and was the route into chaos theory.[26] What Alfred J. Lotka and Vito Volterra did – each working independently and unknown to one another, was to model two or more populations that interacted with each other.

The simplest L-V model is the well-known two-species predator-prey model. There is a logistic equation for each species linked by their interactions. The predator species grows when there is an abundance of prey; the prey species declines when it is eaten by the predator. Not surprisingly there is an oscillating solution. What is more interesting in terms of the translation into other fields is the 'competition for resources' form of the L-V model. In this case, two or more species compete for one or more resources; this provides a way of representing interactions between species in an ecosystem. The translation comes through identifying systems of interest in which populations of other kinds compete for resources. There are examples in chemistry where molecules in a mixture compete for energy, in geography where retailers compete for consumers (as in my own work with Britton Harris in 1978, 'Equilibrium values and dynamics of attractiveness terms in production-constrained spatial-interaction models')[27] and in security with Lewis Fry Richardson's model of arms races and wars.[28] There are undoubtedly many more possibilities.

Lotka, Volterra and Richardson were working in the 1930s and 1940s, and there are interesting common features of their research. None of them worked primarily – at least in the first instance – in ecology. Lotka was a mathematician and chemist, and later an actuary; Volterra was a mathematician and an Italian senator. Both came to mathematical biology relatively late. Richardson was a distinguished meteorologist who later became a college principal. It is worth looking at their original papers to see the extraordinary range of examples they pursued in each case (with real data which must have been difficult to accumulate)

– particularly bearing in mind that there were no computers. Indeed Richardson, at the end of one of his papers, thanks 'the Government Grant Committee of the Royal Society for the loan of a calculating machine'. It was also interesting – though perhaps not surprising, given their mathematical skills – that the men explored the mathematical properties of these systems of equations in various forms and in some depth. Their work at the time was picked up by others, notably V. A. Kostitzin. I picked up a second-hand copy of his 1939 book *Mathematical Biology*[29] via the internet after searching for Volterra's work – Volterra wrote a generous preface for the book.

The Lotka-Volterra equations represent one of the keys to a particular kind of interdisciplinarity, a concept that can be applied across many disciplines because of the nature of what is a generic problem – modelling the 'competition for resources'. In a particular instance of a research challenge, the trick is to be aware that the problem may be generic and that elements of a toolkit may lurk in another discipline.

Exercises

The illustrations in this chapter have been focused on formal models (mathematical models) though it should be borne in mind that the concept is broader – a system diagram constitutes a model, for example. The exercises that follow are based on formal models, but can be interpreted more broadly.

1   The exercises at the end of Chapter 3 should have led you to sketch a model of your system of interest. Does the urban model outlined in above (pp. 38–43) help you to expand it? Or explore an alternative approach.

2   Almost certainly there will be competing models or 'schools of modelling' that are on offer as you try to build your bespoke model for your system of interest. Identify some that may be relevant to your research and write some notes about the principles underpinning them. Can you select a starting point from this set or do you need something really new?

3   Some models have components which can be used as building blocks in other modelling enterprises, such as your research. An example is the 'generalised cost' discussed in this chapter (pp. 48–49), which can be related to utility in economics. Another example is 'accessibility', which turns up as a model by-product, but which can be used to

explore some 'real challenges'. Review the building blocks of your model and see if there are any similar concepts already in use that will help you progress.

4   If you are building a model, examine the core equations and see if any are similar to well-known 'equations with names'. This may provide a route to helpful mathematical analysis from which you can draw.

Notes

1   For a historical overview see Wilson, *Urban Modelling*.
2   Ira S. Lowry, *A Model of Metropolis*.
3   This was a very crude form of economic base model which can be elaborated as an input-output model.
4   There are many examples of Lowry-based models. See, for example, Simmonds, 'The design of the DELTA land-use modelling package', edited and updated in Wilson, *Urban Modelling*, Vol. 5, Chap. 99 and the references therein.
5   See Birkin et al., *Intelligent GIS*.
6   Wilson, 'Boltzmann, Lotka and Volterra and spatial structural evolution'.
7   Dearden and Wilson, 'A framework for exploring urban retail discontinuities'.
8   Lotka, *The Elements of Physical Biology*; Volterra, 'Population growth, equilibria and extinction umder specific breeding conditions'
9   Wilson, 'Boltzmann, Lotka and Volterra and spatial structural evolution'.
10 Wilson and Oulton, 'The corner-shop to supermarket transition in retailing'.
11 First discussed in Chapter 2, p.43.
12 This had one nearer contemporary consequence. I was at the first meeting of the Vice-Chancellors of universities that became the Russell Group. There was a big argument about the name. We were meeting in the Russell Hotel and, after much time had passed, I said something like, 'Why not call it the Russell Group?', citing not just the hotel but also Bertrand Russell as a mark of intellectual respectability. Such is the way in which brands are born.
13 Wilson, 'A general representation for urban and regional models'.
14 See Dearden and Wilson, 'The relationship of dynamic entropy maximising and agent-based approaches', to link spatial interaction and agent-based modelling.
15 Government Office for Science, Foresight, Future of Cities project reports. https://www.gov.uk/government/uploads/system/uploads/attachment_data/file/520963/GS-16-6-future-of-cities-an-overview-of-the-evidence.pdf. Accessed 3 January 2022.
    Science of Cities: https://www.gov.uk/government/uploads/system/uploads/attachment_data/file/516407/gs-16-6-future-cities-science-of-cities.pdf. Accessed 3 January 2022.
    Foresight for Cities: https://www.gov.uk/government/uploads/system/uploads/attachment_data/file/516443/gs-16-5-future-cities-foresight-for-cities.pdf. Accessed 3 January 2022.
    Graduate Mobility: https://www.gov.uk/government/uploads/system/uploads/attachment_data/file/510421/gs-16-4-future-of-cities-graduate-mobility.pdf. Accessed 3 January 2022.
16 Quandt and Baumol, 'The demand for abstract modes'.
17 Lancaster, 'A new approach to consumer theory'.
18 Boyce and Williams, *Forecasting Urban Travel*.
19 Lowry, *A Model of Metropolis*.
20 Stewart, *Seventeen Equations that Changed the World*.
21 Tolstoy, *War and Peace*, p. 915.
22 Jaynes, *Probability Theory*.
23 This story is reported in E. C. McIrvine and M. Tribus 'Energy and information' in *Scientific American*.
24 Hidalgo, *Why Information Grows*.
25 Weaver, 'Science and complexity', cites an interesting example that illustrates the power of Boltzmann. Consider a billiards table with a large number of balls. Launch the white ball and there may be a large number of collisions. The task of solving the equations of motion for each ball

is intractable. However, if we ask a different question, Boltzmann's question, 'What is the average number of times a ball will strike a cushion?', the problem becomes (in principle) tractable.

26 May, 'Stability in multi-species community models'; May, *Stability and Complexity in Model Ecosystems*.
27 Harris and Wilson, 'Equilibrium values and dynamics of attractiveness terms in production-contrained spatial-interaction models'.
28 Richardson, *Arms and Insecurity*.
29 Kostitzin, *Mathematical Biology*.

Chapter 5
Establishing a research base 2: from 'data' to AI

Introduction: *nullius in verba*

The second stage in developing a research project is to explore relevant data sources and particularly to take into account the impact of increased data availability, the so-called age of 'big data'. A following step is to absorb the developments in data science as underpinning technologies for both artificial intelligence (AI) algorithms and their application in a number of domains. The ordering here is deliberate. An argument is sometimes made that analysis of the data will somehow render modelling (and theory?) obsolete. In fact the opposite remains the case – we need the understanding of our systems of interest to know which data are relevant. The systems' focus, as we have argued, is one of the engines of interdisciplinary research. The increased availability and richness of data can now be seen as another, making interdisciplinary approaches more feasible. I examine in turn the development of data science as a framework for the way in which we handle data in our research, then relate this to the possibility of adding AI algorithms to our toolkit (pp. 58–9 and 59–62). I offer an overview of a research future that is data-driven in new ways (pp. 62–7).

Will data science change the world?

The availability of data through a combination of multiple 'sensors' and computing power, so-called 'big data', has the power to revolutionise. It has been described as the 'new oil' in terms of the value it can deliver. Many disciplines are underpinned by mathematics and experimental science is by statistics. Since these are being transformed by data through new disciplines such as machine learning, together with advances in computer science, new foundations are being laid that are essentially interdisciplinary. All of this combines into the field of artificial intelligence. We can spend some time and space, therefore, in exploring these developments.

I have re-badged myself several times in my research career: mathematician, theoretical physicist, economist (of sorts), geographer, city planner, complexity scientist and now, data scientist. Is data science a new enabling discipline (see Chapter 2) or is it firmly located within statistics? A similar question arises for AI. My career path is in part a matter of personal idiosyncrasy, but it is also a reflection of how new interdisciplinary research challenges emerge. I have had the privilege of being the Chief Executive of The Alan Turing Institute, the national institute for data science and AI. Its strapline is 'Data science will change the world'. Data science is clearly the new kid on the block. How has this come about?

First, there is an enormous amount of new 'big' data around; second, this has had a powerful impact on all the sciences; and third, it will exert significant influence on our society, economy and way of life. Data science represents these combinations. The data comes from ubiquitous digitisation combined with the 'open data' initiatives of government and extensive deployment of sensors and devices such as mobile phones. This in turn generates huge research opportunities. In broad terms, data science has two main branches. First, what can we do with the data? (Which includes applications of statistics and machine learning.) Second, how can we transform existing science with this data and these methods? Much of the answer to the second question is rooted in mathematics. To make this work in practice, a time-consuming first step must be taken – the data has to be made useable by combining different sources in different formats, a process known as 'data wrangling'. The whole field of Data science is driven by the power of the computer and computer science, whereas understanding the effects of data on society, and the ethical questions it poses, is led by the social sciences.

All of this combines in the idea of 'artificial intelligence' (AI). In many applications, AI supports human decision-making and the current buzz phrase is 'augmented intelligence', where the 'machine' has not yet passed the Turing test of competing with humans in thought.

I can illustrate the research potential of data science through two examples. The first is drawn from my own field of urban research and the second from medicine. Recent AI research here was learned, no doubt imperfectly, from my Turing colleague Mihaela van der Schaar.

As we have seen, there is a long history of developing mathematical and computer models of cities. Data arrives very slowly for model calibration – the decennial census, for example, is critical. A combination of open government data and real-time flows from, for example, mobile phones and social media networks have changed this situation – real-time calibration is now possible. This potentially transforms both the science and its application in city planning. Machine learning complements, and potentially integrates with, the models. Data science in this case adds to an existing deep knowledge base.

Medical diagnosis is also underpinned by existing knowledge – physiology, cell and molecular biology, for example. It is a skilled business, interpreting symptoms and tests. This can be enhanced through data science techniques – beginning with advances in imaging and visualisation and then the application of machine learning to the variety of evidence available. The clinician can add his or her own judgement (augmented intelligence). Treatment plans follow. At this point, something really new kicks in. Live data on patients, including their responses to treatment, becomes available. This data can be combined with personal data to derive clusters of like patients, enabling the exploration of the effectiveness of different treatment plans for different types of patients. This opens the way to personalised intelligent medicine, set to have a transformative effect on healthcare.

These kinds of developments of the science, and the associated applications, are possible in almost all sectors of industry. It is the role of The Alan Turing Institute to explore both the fundamental science underpinnings and the potential applications of data science across this wide landscape.

The Turing Institute currently works in fields as diverse as digital engineering, defence and security, computer technology and finance, as well as cities and health. This range will expand as this very new institute grows. We will work with and through universities, as well as with commercial, public and third sector partners, to generate and develop the fruits of data science. This is a challenging agenda, but a hugely exciting one.

What is data science? What is AI?

When I first took on the role of CEO at The Alan Turing Institute, the strap line beneath its title was 'The National Institute for Data Science'. A year or so later this became 'The National Institute for Data Science and AI' at a time when there was a mini-debate about whether there should be a separate National Institute for AI. However, it has always seemed clear to me that AI should be included in data science – or maybe vice versa. In the early data science days, plenty of researchers in Turing were focused on machine learning. However, we acquired the new title, for avoidance of doubt, one might say, and it now seems worthwhile to unpick the meanings of these terms. However we choose to define them, there will be overlaps, but we can gain some new insights by making the attempt.[1]

AI has a long history, with well-known 'summers' and 'winters'. Data science is newer and is created from the increases in data that have become available (partly generated by the Internet of Things), closely linked with continuing increases in computing power. For example, in my own field of urban modelling, where we need location data and flow data for model calibration, the advent of mobile phones has provided a data source that locates most of us at any time – even when phones are switched off. In principle, this means that we could have data that would facilitate real-time model calibration. New data, 'big data', is certainly transforming virtually all disciplines, industry and public services.

Not surprisingly, most universities now have data science (or data analytics) centres or institutes, either real or virtual. This has certainly been the fashion, but may now be overtaken by AI in that respect. In Turing, our 'Data science for science' theme has now transmogrified into 'AI for science' as more all-embracing. Some more renaming may follow.

Let us start the unpicking. Big data has certainly invigorated statistics. Indeed, the importance of machine learning within data science is a crucial dimension, particularly as a clustering algorithm with obvious implications for targeted marketing (and electioneering), hence machine learning (and data science) as 'statistics reinvented'. The best guide to AI and its relationship to data science that I have found is Michael I. Jordan's blog post 'Artificial Intelligence – The Revolution Hasn't Happened Yet' – googling the title takes you straight there.[2] He notes that historically AI has stemmed from what he calls 'human-imitative AI', whereas now it mostly refers to the applications of machine learning – 'engineering' rather than mimicking human thought. As this has had huge successes in the business world and beyond, it has come to be called data science

– closer to my own interpretation of data science, but which, as noted, fashion now relabels as AI.

We are a long way from machines that think and reason like humans, but what we have is very powerful. Much of this augments human intelligence and thus, following Jordan, we can reverse the acronym to 'IA' or 'intelligence augmentation'. This is exactly where the Turing Institute works on rapid and precise machine-learning led medical diagnosis, with researchers working hand in hand with clinicians. Jordan also adds another acronym, 'II', meaning 'intelligent infrastructure'. He notes that:

> Such infrastructure is beginning to make its appearance in domains such as transportation, medicine, commerce and finance, with vast implications for individual humans and societies.

This is a larger-scale concept than my notion of an underdeveloped field of research being the design of (real-time) information systems.

This framework, for me, provides a good articulation of what AI means now – IA and II. However, fashion and common usage will demand that we stick to AI. It will most probably be a matter of personal choice whether we continue to distinguish data science within this.

Mix and match: the five pillars of data science and AI

The brief introduction of the previous section indicates that we are in a relatively new interdisciplinary field. It is interesting to continue the exploration by connecting to previous drivers of interdisciplinarity, in order to see how these persist and 'add' to our agenda, and then to examine examples of new interdisciplinary challenges.

There are five pillars of data science and AI, and these help us to develop our discussion. Three make up, in combination, the foundational disciplines of mathematics, statistics and computer science. The fourth is the data, big data as it now is. The fifth is a many-stranded pillar, domain knowledge, that is now combining enabling disciplines with substantive ones. The mathematicians use data to calibrate and test models and theories. The statisticians also calibrate models and seek to infer findings from data. The computer scientists develop the intelligent infrastructure. Above all, the three combine in the development of machine learning – the heart of contemporary AI and its applications. Is this already a new discipline? Not yet, I suspect. It is certainly not distinguished by undergraduate degrees in AI (unlike, say, biochemistry). These four disciplines, as we have noted, can be thought of as *enabling* disciplines, a fact that helps us to unpick the strands of the fifth pillar. Both scientists and engineers are users, as are the applied domains such as medicine, economics and finance, law, transport and so on. As the field develops, the AI and data science knowledge will be internalised in many of these areas, in part serving to meet the Mike Lynch challenge by incorporating prior knowledge into machine learning.[3]

It has been argued in earlier sections that the concept of a *system of interest* drives interdisciplinarity. This is very much the case here, in the domains for which the AI toolkit is now valuable. More recently, *complexity science* was acknowledged to be an important driver, with challenges articulated through Weaver's notion of 'systems of organised complexity'. This emphasises both the high dimensionality of systems of interest and the nonlinear dynamics that drive their evolution. There are challenges here for the applications of AI in various domains. Handling big data also itself drives us towards high dimensionality. As noted in Chapter 4, I once estimated the number of variables I would like to have to describe a city of one million people at a relatively coarse grain – the answer came out as 10^{13}. This raises new challenges for the topologists within mathematics – how to identify structures within the corresponding data sets, a very sophisticated form of clustering in very high-dimensional state space.

These kinds of system can be described through conditional probability distributions, again with large numbers of variables which pose high-dimensional challenges for Bayesian statisticians. One way to proceed with mathematical models that are high-dimensional and hence intractable is to run them as simulations. The outputs of these models can then be treated as data. There is, to my knowledge, an as yet untouched research challenge: to apply unsupervised machine learning algorithms to these outputs to identify structures in a high-dimensional non-linear space.

We begin to reveal many research challenges across both foundational, and especially, applied domains. In fact we may conjecture that the most interesting foundational challenges emerge from these domains. We can then make another connection, following on from pp. 20–22, to Brian Arthur's argument in *The Nature of Technology*.[4] A discovery in one domain can, sometimes following a long period, be transferred into other domains. These are opportunities we should look out for.

Is it possible to optimise how we do research in data science and AI? We have starting points in the ideas of systems analysis and complexity science – define a system of interest and recognise the challenges of complexity. Seek the data to contribute to scientific and applied challenges (not the other way round) and that will lead to new opportunities?

Perhaps above all, however, seek to build teams that combine the skills of mathematics, statistics and computer science, integrated through both systems and methods foci. This is non-trivial, not least due to the shortage of these skills. In the projects in the Turing Institute funded by the UKRI Special Priorities Fund, *AI for Science and Government* (ASG) and *Living with Machines* (LWM), we are trying to do just this. It is early days, however, and the effectiveness of particular skill mixes is yet to be tested.[5]

Big data and high-speed analytics

My first experience of big data and high-speed analytics was at CERN and the Rutherford Lab over 50 years ago. I was in the Rutherford Lab, part of a large, distributed team working on a CERN bubble chamber experiment. There was a proton-proton collision every second or so which, for the charged particles, produced curved tracks in the chamber. These were photographed from three different angles. The data from these tracks was recorded in something called the Hough-Powell device (named after its inventors) in real time. This data was then turned into geometry, which was then passed to my program. I was at the end of the chain. My job was to take the geometry and work out for this collision which of a number of possible events it actually was – the so-called kinematics analysis. This was done by chi-squared testing, which seemed remarkably effective. The statistics of many events could then be computed, hopefully leading to the discovery of new (and anticipated) particles – in our case the Ω^-. In principle, the whole process for each event, through to identification, could be done in real time, though in practice my part was done offline. This was in the early days of big computers – in our case the IBM 7094. I suspect now it will be all done in real time. Interestingly, in a diary I kept at the time, I recorded my then immediate boss, John Burren, as remarking that, 'We could do this for the economy, you know.'

So if we could do it then for quite a complicated problem, why do we not do this now? Even well-known and well-developed models – transport and retail, for example – typically take months to calibrate, usually from a data set that refers to a single point in time. We are progressing to a position at which, for these models, we could have the data base continually updated from data flowing from sensors. (There is an intermediate processing point, of course – to convert the sensor data into what is needed for model calibration.) This should be a feasible research challenge. What would have to be done? I guess the first step would be to establish data protocols, so by the time the real data reached the model – the analytics platform – it was in some standard form. The concept of a platform is critical here, as it would enable the user to select the analytical toolkit needed for a particular application. This could incorporate a whole spectrum, ranging from maps and other visualisation to the most sophisticated models – static and dynamic.

There are two possible guiding principles for the development of this kind of system: a) what is needed for the advance of the science and b) what is needed for urban planning and policy development. In either case we would start from an analysis of 'need' and thus evaluate what is available from the big data shopping list for a particular set of purposes – probably quite a small subset. There is a lesson here: to think about what we need data for rather than taking the items on the shopping list and asking what we can use them for.

Where do we start? The data requirements of various analytics procedures are pretty well known. There will be additions, for example, incorporating new kinds of interaction from the Internet of Things world.

So why do we not do all this now? Essentially because the starting point – the first demo – is a big team job and no funding council has been able to tackle something on this scale. Therein lies a major challenge. Perhaps we need, as I once titled a newspaper article 'A CERN for the social sciences.' See Chapters 9 and 10 below.

A data-driven future?

We can now build on the argument of the two previous sections and look ahead in more detail. A data-driven transformation is in progress and the infrastructure is being put in place. Infrastructure serves purposes – transport systems provide mobility and accessibility and utilities provide energy and water. The purposes cannot be served without the infrastructure. Much of how we now live and how the economy functions is now driven by data. The flow of data is a new utility and so the provision of the corresponding infrastructure is critical. This provides the means of collecting data, storing it, moving it around to the points of use. To do this requires smart administration, sensors, computer storage, fibre and wireless communications networks and computer power, analytics and display. The key point, as with other infrastructure, is to work backwards along this sequence – from the uses of the data to the articulation of the infrastructure.

We live in the 'information age' – the fourth industrial revolution. Information is coded as data, which once upon a time would mean through the printed word in books and newspapers and financial accounts handwritten into ledgers. The key to the new 'age' is the digitisation of all kinds of data, computer power and the World Wide Web and the internet.

This all has a long history. It was Claude Shannon, in his famous 1945 paper on 'the mathematics of information',[6] who demonstrated the power of the 'binary bit' and provided the basis for the transmission of data. Computing power and the internet have already transformed how businesses and services work, and how we live. Administration, manufacturing, investment and planning, sales and delivery have all changed. Note the increase in the activities of white vans and the ubiquity of the 'Deliveroos'.

But there is more to come. The newer fields of 'data science' and 'artificial intelligence' will be transformative in new ways. What has been achieved so far is largely in specialised silos. The challenge is to explore the futures of the big systems in business and the professions, and in science, engineering, health, education and security and to consider how these combine in the places where most of us live – in cities or city regions. If we take the growth of computing power and the continued expansion of communications bandwidth as given, there are two parts of data science that underpin future developments in applied domains – data wrangling and data analytics.

Data systems are messy. Take one element of one example: the data assigned to a patient which is the input for clinical judgement and for diagnosis. This will range from historic handwritten 'doctor's notes' through all the usual tests, to imaging through MRI scans. Data wrangling is the task of digitising all of this (where necessary) and then making it available in common formats that can be inputs to the analytics. In this medical example, this organised data system can be input to machine learning algorithms and indeed combined with social data on the patient to provide augmented intelligence for the clinician. There are in fact already examples where the algorithm may provide a quicker or better diagnosis. Messy data makes data wrangling an untidy and time-consuming process, but it is a necessary one. It is estimated that data wrangling can take 80 per cent of the time in developing applications. A key research challenge, therefore, is to find a way of automating this.

Data analytics can be thought of as a toolkit. The appropriate tools can be deployed on the data to fulfil the purpose of the application. In some cases, this may be a matter of traditional statistical analysis. In other cases, there may be significant prior knowledge which has been encoded in mathematical models on computers; this can be used to explore alternative plans and futures on the computer rather than through what might turn out to be very poor investment. Such a situation is commonplace in transport planning and retail store location, for example. However, these skills have also been combined into the methods of machine learning (ML). ML algorithms can be applied to data sets to 'learn' routine business processes – for example, to be able to recommend whether to accept a person for insurance (and to estimate the premium) without human intervention. Retailers can classify us in ways that allow them to target their marketing through 'recommender systems'. ML-based computers become learning machines that can replace, even enhance, human judgement in a range of situations. Data analytics provides the foundations of artificial intelligence. There are dangers, of course – serious ones, including bias. In most cases involving automated or semi-automated decision-making that impacts directly on people, checks are either built into the software or, perhaps more commonly, there is some kind of appeals procedure which refers the decision back to a case manager. A related and important research area is the need to make algorithmic outputs interpretable or explainable, particularly difficult in the case of 'deep learning' examples.

We can now return to the central questions. What are the uses of data within a range of application domains and what does this imply for the future development of data infrastructure? The health service provides a good example of what the future could, or will, look like. We can build a scenario. A patient presents, data are collected via tests and/or imaging, then the clinician makes a diagnosis and specifies a treatment plan. Now suppose all this is fully digitised and recorded and the impacts of treatment plans are tracked over time and evaluated. Add machine learning at the diagnostic stage, combining medical and social data. Over a large sample, patients can be clustered by social 'type', disease and stage of disease. In each case the effectiveness of the evaluated treatment plan can be recorded. With this system in place, the clinician has augmented intelligence when a new patient presents and he or she can then select the best treatment plan from the past experiences of thousands, indeed millions, of previous patients.[7] Personalised and super-effective medicine thus becomes achievable. This is feasible, but the inhibiting factor is the availability of data infrastructure. Past and often-failed attempts to digitise medical records demonstrate the scale of the challenge.

The illustration is paradigmatic – the structure can be applied to any system that is client- or consumer-driven such as tracking students through the education system, personalised finance or the progress of offenders through the criminal justice system. Businesses will create their own data infrastructure. Retailers, as noted, are already well down this path. Advanced manufacturers are linking data-driven demand analysis with robotic production. However, all of these gains will be in silos. The big systems are all interdependent. Consider the agenda of the National Infrastructure Commission: transport, utilities, telecoms and broadband and even housing.[8] There are huge opportunities for 'smart' efficiencies, but they demand linking of the major data infrastructure systems. The Commission has plans for a 'digital twin' of this huge linked system, and this would provide a map of the substantive underlying data infrastructure.

The core of the interdependence lies in the functioning of cities, which provide an example of how we have to integrate to capture the relevant complexity. Each city will aim to be 'smart', but data-driven analytics can also contribute to the most challenging economic question: how to improve productivity outside the Greater South East region? To tackle this question, we must turn to the key processes of government – planning and investment, both public and private. What are the best investment strategies to manage projected, and substantial, population growth? How can urban economies develop both in relation to productivity and the 'hollowing out' challenge? How can services be delivered with decreasing budgets? Investment in utilities and transport? What are the implications for urban form? Higher densities and redefinition of green belt areas? How can private and public investment be integrated to best effect? We should also consider an overriding and neglected question: how can all of this be done in a sustainable, carbon-reducing way? Governance and planning structures are needed that respond to this agenda.[9] The data infrastructure systems are key.

We also need better data analytics. This sits 'on top of' the data infrastructure so that future scenarios can be explored and paths towards good 'plans' charted in terms of investment and the allocation of scarce resources. We have the capability to do this, but this is not being deployed systematically – indeed barely at all. What are the barriers and how can we overcome them? Most cities have neither good data technologies nor good analytics and the challenge is to provide an effective capability to local authorities across the country – 164 in England alone. The current 'supply' is very fragmented and the demand is very weak, although the biggest players such as London and Manchester have at least the beginnings of something good, as demonstrated in the London Infrastructure 2050 study. The big players in engineering are well-equipped – Atkins, Arup, IBM and Siemens, for example – a substantial list. Yet they are playing in a weak market in the UK and perhaps doing better as exporters. Is there a market failure here? Should the government be taking a stronger lead?

The National Infrastructure Commission will be very important. The Alan Turing Institute, as the national institute for data science, can also play an important role. It has an urban analytics theme and indeed all its themes contribute to the development of the 'national data infrastructure'.

A second integrator is illustrated by the industrial strategy of the UK government. The challenges of providing the data science base for industrial development are explicitly recognised, particularly through research needs, but also through skills and infrastructure in pillars 1–3 of the strategy. The need for adequate human capital and the corresponding training needs are urgent. Skills are needed at all levels, from the basics of coding data to post-doctoral research. In the Edinburgh region alone, it has been estimated that 10,000 additional data scientists per annum will be needed over the next 10 years. Master's courses in data science are being developed in universities all over the country. Since the field is developing so rapidly, these courses must be integrated with research and industry experience, along with the task of ensuring international competitiveness.

A third cross-cutting theme is data ethics. Questions range from ensuring privacy and anonymity where appropriate and necessary to being able to demonstrate the transparency of machine learning algorithms. In terms of privacy, there is also a prior issue – ensuring the security of the cloud against hacking – and indeed this extends into the wider and rapidly developing field of cyber security. The effectiveness of artificial intelligence and data science in many fields depends on good data being loaded on to the infrastructure. Much of this is personal data, which can in principle be anonymised. This is a non-trivial exercise. A file of such data can often be cross-referenced to publicly available personal data in such a way that identities are revealed. This should be a solvable problem, but it also needs to be convincingly recognised as 'solved'. The transparency issue is mathematically challenging. A company or a government department might use a 'deep learning' algorithm, involving many layers of a neural network, to generate a decision on, say, an insurance policy or a benefits recommendation. At present it is often not possible to give an explicit account of how the decision was reached thus it is difficult to respond to an appeal against the decision. These are major issues and the Nuffield Foundation, the Royal Society, the Royal Statistical Society and The Alan Turing Institute formed a partnership to ensure that progress is made, which has led to the creation of the Ada Lovelace Institute.[10]

In summary, the future will be data-driven. Lives and economies will be transformed by data, data science and artificial intelligence. There are challenges in data wrangling as a starting point and in artificial intelligence as an end point. The future is already evident in sophisticated applications in sectors such as retailing. There are the beginnings in a public service sector such as health, but in that case the ultimate benefits will come from a system-wide application. The industrial sectors are mixed in performance. Robotics-based advanced manufacturers are, unsurprisingly, the leaders and there are beginnings in financial services while others have yet to start. The challenges are interdependent and this is recognised in the thinking about cities and in the industrial strategy. There is a strong argument that the ethics agenda should keep pace with the development of our data-driven futures. However, none of this will happen without effective data infrastructure.

Exercises

'Nullius in verba' is the motto of the Royal Society. It roughly translates as 'Do not take anybody's word for it'. In other words, data is crucial for theory (or hypothesis) and model testing. We live in an age where there is an abundance of data and a developing discipline of 'data science', which offers tools for its management and associated analytics, whether statistics, machine learning or, more broadly, AI. This is, therefore, a crucial part of developing a research project.

1   Identify possible data sources for your evolving research project.
2   Explore ways of handling the challenge of 'missing' data – data you would like to have, perhaps data that is essential. What are the best ways of estimating it from available data? Explore whether there are submodels that will help with this. Consider methods such as biproportional fitting which help when there is partial data.
3   Explore the data science and AI literature to find possible new tools.

Notes

1   Hall and Pesenti, *Growing the Artificial Intelligence Industry in the UK*.
2   Jordan, M. I., 'Artificial intelligence: the revolution hasn't happened yet'.
3   Mike Lynch gave a 'history of AI' talk to a Turing conference. He ended by arguing that the 'incorporation of prior learning' was the biggest future challenge for AI.
4   Brian Arthur, *The Nature of Technology*.
5   See ASG in www.turing.ac.uk.
6   Claude Shannon, 'A Mathematical Theory of Cryptography', classified memorandum for Bell Telephone Labs, 1945; see also Shannon and Weaver, *The Mathematical Theory of Communication*.

7   This analysis is based on the work of Professor Mihaela van der Schaar of The Alan Turing Institute and the University of Cambridge.
8   National Infrastructure Commission, *Data for the Public Good*.
9   For a detailed account of the challenges see the Government Office for Science, Foresight Future of Cities (2013) and the associated websites.
10 See The Ada Lovelace Institute website, adalovelaceinstitute.org.

Chapter 6
Doing the research: different kinds of problem solving

Introduction

Through Chapters 3, 4 and 5, it should have been possible to at least sketch out a research area. The next task is to develop this thinking into a research problem to be tackled. The starting point is clearly the science: is the system of interest sufficiently well understood and researched to move straight on to real challenges? In this chapter, therefore, we begin with the science (pp. 69–70) and move on to consider the application of the science to real challenges. We explore the level of ambition needed to take on challenges from the 'wicked problems' list (pp. 70–3) and then look at the practical tasks associated with engaging with public policy (pp. 70–4). We illustrate the argument with an example: the future of cities (pp. 72–3) before introducing a different way of tackling real challenges – working with industry (pp. 75–7).

Science challenges

All sciences have research challenges. The STM framework can be used as a way of exploring these. In the physical and biological sciences, systems can be more or less isolated for research purposes, which helps. In the social sciences, however, our systems of interest are living and evolving in ways beyond Darwin's biology. In one sense at least, the systems are more complex. (Perhaps this is the answer to the question 'what would Warren Weaver say now?' which we take up in Chapter 7.) One aspect of this complexity is the fact that in seeking to understand the evolution of cities, for example, some key elements turn on individuals making decisions about, for example, land development.[1] Thus, while we can model transport systems pretty well for the here and now, the modelling task becomes much more difficult if we are seeking to model the evolution of urban form.

The STM framework leads us into an analysis of research challenges on all three dimensions. In the urban context, for example, an important part of system definition is the specification of variables which represent the urban economy – and the associated model would be vital for policy development. However, this submodel is inadequate, not least because of lack of data on, for example, inter-city trade. This weakness may be one reason why urban models are not ubiquitous within urban planning (in its broadest sense). Omnipresent is the issue of scale, from individuals and organisations through to macroscales involving cities, regions and countries. Theorising and modelling must take place at each of these scales and, perhaps most importantly, connecting across them.

We have to make judgements continually about attempting to fill these kinds of gaps. In so doing we must define and possibly extend the system definition to identify research challenges. At the same time we must look for weaknesses in the current theory – the knowledge base. Are the methods adequate? Can we search the methods' space more widely, possibly taking in what is new, such as AI?

Real challenges: ambition and 'wicked problems'

Real challenges, as we have seen above (pp. 6–11), pose some very difficult problems which can be classified as 'wicked'. There are many definitions of wicked problems, first characterised by Rittel and Webber in the 1970s.[2] Essentially, however, they are problems that are well known and difficult, ones that governments of all colours have attempted to solve.

My own list, relating to cities in the UK, would be something like:
- Social
  - Social disparities, welfare, unemployment, pensions and housing.
- Services   ✔
  - Health (elements of postcode lottery and poor performance)
  - Education (a long tail of poor performance, for both individuals and schools)

- Prisons (high levels of recidivism).
- Economic
  - Productivity outside the Greater South East area)
  - 'Poor' towns – seaside towns, for example
  - Global, with local impacts, for example, sustainability (responding to the globalisation of the economy and climate change)
  - Food security, energy security and indeed security more broadly.

There are lots of ways of exploring this and much more detail could be added. See, for example, *The Too Difficult Box*, edited by Charles Clarke.[3]

Even at this broad level of presentation, the issues all connect and this is one of the arguments, continually put, for joined-up government. It is almost certainly the case, for example, that the social challenges have to be tackled through the education system. Stating this is insufficient, however. Children from deprived families arrive at school relatively ill-prepared – in terms of vocabulary, for example – and so start, it has been estimated, two years 'behind'. In a conventional system, there is a good chance that they may never catch up. There are extremes of this as noted: children who have been in care, for example, rarely progress to higher education; it then turns out that quite a high percentage of the prison population have at some stage in their lives been in care. Something fundamental is wrong there.

We can conjecture that there is another chain of causal links associated with housing issues. Consider not the overall numbers issue – that in the UK we build 150,000 houses a year when the 'need' is estimated at 300,000 or more – but the fact that there are areas of very poor housing, usually associated with deprived families. I would argue that this is not a housing problem but an income problem – not enough resource for the families to maintain the housing. It is an income problem because it is an employment problem. It is an employment problem because it is a skills problem. It is a skills problem because it is an education problem. Hence the root, as implied earlier, lies in education. A first step in seeking to tackle the issues on the list is thus to identify the causal chain and to begin with the roots.

How then can we take a 'sledgehammer' to these challenges? If the problem can be articulated and analysed, then it should be possible to see 'what can be done about it'. The investigation of feasibility then kicks in, of course. Solutions are usually expensive. However, we do not usually manage to do the cost-benefit analysis at a broad scale. If we could be more effective in providing education for children in care, and in achieving the rehabilitation of more prisoners, expensive schemes could be paid for by savings in the welfare and prison budgets – governments must invest to save.

Let us start with education. There are some successful schools in deprived areas, so examples are available. (This may not pick up the childcare issues, but we return to those later.) Many studies maintain that the critical factor in education is the quality of the teachers, so enhancing the status of the teaching profession and building on schemes such as Teach First will be very important. Much is being done, but not enough. Above all, there must be a way of refusing to accept failure in any individual cases. With contemporary technology, tracking is surely feasible, though the follow-up might involve lots of one-to-one work, which is expensive. Finally, there is a legacy issue – those from earlier cohorts who have been failed by the system may be part of the current welfare and unemployment challenge. Again, some kind of tracking, some joining-up of social services, employment services and education, should provide strong incentives to engage in lifelong learning programmes. Serious catching up is needed here. The tracking and joining-up part of this programme should also deal with children in care as a special case, while a specific component of the legacy programme should focus on the prison education and rehabilitation agenda.

There is then an important add-on – it may be necessary for the state to provide employment in some cases. Consider people who have spent time in prison. They are potentially unattractive to employers (though some are creative in this respect) and so obtaining employment through, let's say, a Remploy type of scheme[4] – maybe as a licence condition of (early?) release – becomes a partial solution. This might help to take the UK back down the league table of prison population per capita. This could all in principle be done and paid for out of savings – though there may be an element of 'no pain, no gain' at the start. There are examples across the UK of where it is being done. Let us see how these could be scaled up.

Similar analyses could be brought to bear on other issues. Housing is at the moment driven by builders' and developers' business models and needs to be tackled in these terms. As with teacher supply, there is a skills capacity issue. Any initiatives need to be underpinned by education, employment and welfare reforms, along with contributions by planners who can switch from a development control agenda to a place-making one.

In all cases radical thinking is required, but realistic solutions are available. We can offer a Michael Barber checklist for tackling problems (see his 2016 book, *How to Run a Government so that citizens benefit and taxpayers don't go crazy*[5]) which is very delivery-focused, as is his wont. In solving a particular problem the following questions must be asked: what are you trying to do? How are you going to do it? How do you know you will be on track? If you are not on track, what will you do?

This exploration can be illustrated in more depth through the publications of the Government Office for Science Foresight Project,

The Future of Cities.[6] It has 'reported', not in the conventional way, with one large report and many recommendations, but with four reports and a mass of supporting papers. Googling 'Foresight Future of Cities' leads very quickly to the website and all the supporting material.[7] The project engaged with 14 government departments – 'cities' as a topic crosses government – and over 20 cities in the UK were visited and consulted.

The key point that the problems and challenges are all interdependent was reinforced through this study, which places substantial responsibility on researchers, analysts and policy makers to handle this and not to work in silos. If the project made one key recommendation, that was it. Beyond that, it was recognised that there were no easy solutions, but that it was possible to formalise the process of generating and exploring alternative scenarios – the subject of one of the reports. This is the basis for inventing possible plans –the 'design' part of 'PDA'.

Research into policy: lowering the bar

The task of working with government on policy by relating research and real challenges is non-trivial. Some time ago I attended a British Academy workshop on 'Urban Futures' – partly focused on research priorities and partly more specifically on research that would be useful for policy makers. The group consisted mainly of academics who were keen to discuss the most difficult research challenges. I found myself sitting next to Richard Sennett[8] – a pleasure and a privilege in itself, as he was someone I had read and knew by repute, but had never met. When the discussion turned to research contributions to policy, Richard made a remark which resonated strongly with me; it was to make the day very much worthwhile. He said, 'If you want to have an impact on policy, you have to lower the bar.' We discussed this briefly at the end of the meeting, and I hope he will not mind if I try to unpick it a little. It does not tell the whole story of the challenge of engaging the academic community in policy, but it does offer some insights.

The most advanced research is likely to be incomplete and to have many associated uncertainties when translated into practice. This can offer insights, but the uncertainties are often uncomfortable for policy makers. If we lower the bar to something like 'best practice', this may involve writing papers and producing presentations which do not offer the highest levels of esteem in the academic community. What is on offer to policy makers has to be intelligible, convincing and useful. Being convincing means that what we are describing should be evidence-based.

And when these criteria are met, of course, there should be another kind of esteem associated with the 'research for policy' agenda. I guess this is what 'impact' is supposed to be about (though I think that is only half of the story, since impact that transforms a discipline may be more important in the long run).

'Research for policy' is, of course, 'applied research'. This also brings up the esteem argument that if research is 'applied', then it is deemed less 'esteemful', if I can make up a word. In my own experience, engagement with real challenges – whether commercial or public – adds seriously to basic research in two ways. First, it throws up new problems. Second, it provides access to data – for testing and further model development – that simply would not be available otherwise. Some of the new problems may be more challenging and, in a scientific sense, more important than the old ones.

So, we return to the old problem: what can we do to enhance academic participation in policy development? First let me offer a warning: recall the policy-design-analysis argument introduced in Chapter 1. Policy is about what we are trying to achieve, design is about inventing solutions and analysis is about exploring the consequences of, and evaluating, alternative policies, solutions and plans. The point here is that analysis alone, the stuff of academic life, will not of itself solve problems. Engagement, therefore, ideally means engagement across all three areas, not just analysis.

How can we then make ourselves more effective by lowering the bar? First, we should ensure that our 'best practice' is intelligible, convincing and useful: evidence-based. This means being confident about what we know and can offer. However, we also need to be open about what we do not know. In some cases we may be able to say that we can tackle, perhaps reasonably quickly, some of the important 'not known' questions through research; and that may need resource. Let me illustrate this with retail modelling. We can be pretty confident about estimating revenues (or people) attracted to facilities when something changes – a new store, a new hospital or whatever – even considering the impact of internet retailing or 'virtual' medical consultations. Then there is a category, in this case, of what we 'half know'. We have an understanding of retail structural dynamics, to a point where we can estimate the minimum size that a new development has to be for it to succeed. But we cannot yet do this with confidence. So a talk on retail dynamics to commercial directors may be 'above the bar'.

I suppose another way of putting this argument is that for policy engagement purposes, we should know where to set the height of the bar – confidence below, uncertainty (possibly with some insights) above. A whole set of essays remain to be written on this for different possible application areas.

Spinning out

I estimate that once every two years for the last 20 or 30 years there has been a report of an inquiry into the effectiveness of the transfer of university research into the economy – for commercial or public benefits. This is a version of 'research for' rather than just 'research on' that we discussed above (pp. 33–5). The fact that the sequence continues demonstrates that this remains a challenge. One mechanism is the spinning out of companies from universities and this section is in two parts: the first describing my own experience and the second seeking to draw some broader conclusions. Either part may offer some clues for budding entrepreneurs. What is demonstrated here is that this kind of approach to what amounts to applied 'research for' has to be interdisciplinary.

This is the story of GMAP Ltd. The idea was born, half-formed, at Wetherby Racecourse on Boxing Day 1984 and it was six years before the company was spun out. Of course, there is a back story. Through the 1970s I worked on a variety of aspects of urban modelling supported by large research grants. All of this work was basic science, but there was a clear route into application, particularly into town planning. By the early 1980s the research grants had dried up – a combination of becoming unfashionable and a general feeling that perhaps it was 'someone else's turn'. I worked with a friend and colleague, Martin Clarke, and he and I always went to Wetherby races on Boxing Day. There we discussed the falling away of research grants. As we watched – on television in a bar – Borough Hill Lad win the King George VI Stakes at Kempton, we somehow decided that the commercial world would provide an alternative source of funding. We had models at our disposal – notably the retail model. Surely there was a substantial market for this expertise.

Our first thought was that the companies with the resources to implement this idea were the big management consultants. In 1985 we began a tour of possible candidates. The idea was that they would work with our models on some kind of licence basis. Martin and I could be consultants and they would find the clients. We were well received and usually given a good lunch – then nothing happened. It became clear that DIY was the only way to make progress. We approached some companies whom we knew and thought were possible targets, but mostly our marketing was cold calling, based on a weekly reading of *The Sunday Times* job advertisements to identify companies seeking to fill marketing posts. Over two years we gained a number of small contracts, run through ULIS (University of Leeds Industrial Services). We learned our first lesson in this period – to get contracts, we had to provide the information that the companies actually wanted rather than what we thought they should have. We thought we could offer the Post Office a means of optimising their network; what they actually wanted was to know the average length of a garden path. That was our first contract and that is what we did. Another company (slightly later) said that all these models were very interesting, but their data was in 14 different information systems – could we sort that out? We did. The modelling came later.

Our annual turnover in Year 1 was around £20k and it slowly grew to around £100k. The big breakthroughs came in 1986 and 1987, when we won contracts with WHSmith and with Toyota. By then we had our first employee and GMAP became a formal division of ULIS. It was not yet spun out, but we could run it like a small company, with shadow accounts that looked like real ones. There was then steady growth and in late 1989 we won a major contract with the Ford Motor Company. By 1990 our annual turnover had reached £1 million and we had a staff of around 20. We crossed a threshold and were allowed to spin out as GMAP Ltd. By this time I was heavily involved in university management and so could function only as a non-executive director. The company's development turned critically on Martin becoming the full-time managing director.

The 1990s were years of rapid growth. We retained clients such as WHSmith, Toyota and Ford, but added BP, SmithKline Beecham, the Halifax Building Society and many more. We were optimising retail, bank and dealership networks across the UK and, in the case of Ford, all over Europe. By 1997 our turnover was almost £6 million and we were employing 110 staff. And then came a kind of ending. In 1997 the automotive part of GMAP was sold to R.L. Polk & Company, an American company, and the rest, in 2001, to the Skipton Building Society, to merge into a group of marketing companies that they were building.

What can we learn from this? It was very hard work, especially in the early days. DIY meant just that – Martin and I wrote the computer programs, wrote and copied the reports and collected the data. We once stood outside Marks & Spencer in Leeds with clipboards asking people where they had travelled from so that we could get the data to calibrate a model. We were moved on by the shop's staff for being a nuisance. We also had to be very professional. A project could not be treated like a conventional research project. We had to learn how to function in the commercial world very quickly – if there was a three-month deadline, it had to be met. But it was exciting as well. We grew continuously. We did not need any initial capital but funded ourselves out of contracts. We were always profitable. It was real research – we had incredible access to data from companies that would have been unavailable if we had not been working for them. And many of them rather liked being referred to in papers published in academic journals.

Could it be done again? In this field, possibly, though this kind of analysis has become more routine and has been internalised by many – notably the big supermarket companies. However, there are many companies that could use this technology with (literally) profit, but do not. And there are huge opportunities in the public sector, notably education and health. The companies we worked with, especially those with whom we had long-term relationships, recognised the value of what they were getting – it impacted on their bottom line. We did relatively little work in the public sector – not for want of trying, but it proved difficult to convince senior management of the value. However, it could certainly be done again on the back of new opportunities. Much is said about the potential value of big data or of the Internet of Things, for example, and many small companies are now in the business of seeking out new opportunities. But is anyone linking serious modelling with these fields? Now there's an opportunity.[9]

Exercises

1   Review the science challenges around your system on interest.
2   Decide how would you frame a description of your research to a policy maker (who was not a researcher and did not have extensive analytical skills).
3   Consider the commercial and industry sectors (in the broadest sense, possibly including management consultants, for example) and explore how the results of your research might be of use in one or more of those sectors. Write a prospectus for establishing a spin-out company.

Notes

1   This takes us back to the distinction between exogenous and endogenous variables in models.
2   Rittel and Webber, 'Dilemmas in a general theory of planning'.
3   Clarke,C (ed.), *The Too Difficult Box*.
4   Remploy was a government-funded company that provided employment for disabled people. It would not be difficult to develop such a concept in relation to ex-offenders, the costs of which would almost certainly be more than met by savings.
5   Barber, *How to Run a Government*.
6   I chaired the expert group which led this study.
7   Government Office for Science, The Future of Cities project reports.
8   See, for example, Sennett, *The Corrosion of Character*; *The Culture of the New Capitalism*; *The Craftsman*.
9   See Clarke, M, *How Geography Changed the World and My Small Part in It* for a detailed history.

Part 3
Tricks of the trade

Chapter 7
Adding to the toolkit 1: explorations

Introduction

In Parts 1 and 2, we have established the elements that enable us to define research challenges. In this part, covering Chapters 7 and 8, we explore ways of adding to our research toolkit and increasing our 'tricks of the trade'. In this chapter we first consider the importance of understanding the history of a field so that we can identify the game-changers and how we can mine the past (pp. 82–4 and pp. 84–6). We then examine to what extent current research is determined by fashion and seek to understand the implications of this (pp. 86–9). It is sometimes possible to identify interesting research problems by transferring concepts and methods that have been developed in one context into another discipline. An example in my own experience connects urban modelling of the contemporary to historical contexts (pp. 89–91). Finally we address a different kind of question: what would Warren Weaver do now? (pp. 92–4). Weaver was the Science Vice-President of the Rockefeller Foundation in the 1950s. Such a perspective offers a different approach to the question of where the funders of research should be spending their money.

Learning from history

My core research career was built on my move from the Rutherford Laboratory at Harwell to the Institute of Economics and Statistics in Oxford. I was recruited in the autumn of 1964 by Christopher Foster (now Sir Christopher) to work on the cost-benefit analysis of major transport projects. My job was to do the computing and mathematics and at the same time to learn some economics. The project needed good transport models and at the time all the experience was in the United States. Christopher had worked with Michael Beesley (LSE) on the pioneering cost-benefit analysis of the Victoria Line.[1] To move forward on modelling, Christopher, Michael and I embarked on a tour of the US in 1965. As I remember, in about 10 days we visited Santa Monica and Berkeley, Philadelphia, Boston and Washington, D.C. We met a good proportion of the founding fathers – they were all men – of urban modelling. A number of them influenced my thinking in ways that have been part of my intellectual make-up ever since – threads that can easily be traced in my work over the years. An interesting question thus arises: for those recruited in subsequent decades, and indeed in the present and near-past, what are the equivalents? It would be an interesting way of both writing a history of the field and of looking for new routes forward.

Jack (I. S.) Lowry was working for the RAND Corporation in Santa Monica where he developed the model of Pittsburgh that now bears his name – outlined in Chapter 4 (pp. 38–43). I recall an excellent dinner in his house overlooking the Bay. Lowry's model has become iconic because it revealed the bare bones of a comprehensive model in the simplest way possible. Those of us involved in building comprehensive models have been elaborating it ever since. For me the conversation reinforced something that was already becoming clear – the transport model needed to be embedded in a more comprehensive model so that the transport impact on land use (and vice-versa) could be incorporated.

The second key proponent of the comprehensive model was Britton Harris, Professor of City Planning at the University of Pennsylvania. Particularly important in the context of that visit, however, was his role as Director of the Penn Jersey Land-Use Transportation Study. The title indicated its ambitions. This again reinforced the 'comprehensive' argument and our meeting became the basis of a life-long friendship and collaboration. I was to spend many happy hours with Brit and his wife Ruth in their house in Wissahiken Avenue in Philadelphia. The Penn Jersey study used a variety of modelling techniques, not least mathematical programming, a new element of my intellectual toolkit. More of Brit later. At Penn – though I am not sure now whether it was on the same trip or later – I met Walter Isard, a giant figure in the creation of regional science. He contributed to my roots in regional input–output modelling. Walter was probably the first person to recognise that von Thunen's theory of rent could be applied to cities – see Isard's book *Location and the Space-Economy*, published in 1956.[2] Bill Alonso, one of his graduate students, fully developed the theory of bid-rent. We visited Bill in Berkeley and I recall a letter from him three years later, in the heady days of 1968, which began, 'As I write, military helicopters hover overhead'. At Penn, it was Ben Stevens who operationalised the Alonso model in his 1960 paper with John Herbert as a mathematical programming model.[3] This fed directly into work I did in the 1970s with Martyn Senior to produce an entropy-maximising version. This made me realise that one of the unheralded advantages of the entropy method was its ability to make optimising economic models – such as the Alonso-Herbert-Stevens model – 'optimally blurred' to acknowledge the sub-optimal real life.[4]

In Harvard we met John Kain. Very much the economist, he was very concerned with housing models, territory I have failed to follow up on since. He was at the Harvard-MIT Joint Centre for Urban Studies whose existence was a sign that these kinds of interdisciplinary centres were fashionable at the time. They have been in and out of fashion ever since, but are fortunately now fashionable again. An alumnus was Martin Meyerson, by this time Chancellor of the University of California at Berkeley. We dined with him in his rather austere but grand official dining room – why does one remember these things rather than the conversation? There also was Daniel (Pat) Moynihan who had just left the Centre to work for the President in Washington – another sign of the importance of the urban agenda. I was urged to meet him and that led to my only visit to the White House – to a small office in the basement. He later became very grand as a long-serving Senator for New York State.

The Washington part of our visit established some other important contacts and building bricks. We engaged directly with the transport modelling industry through Alan Voorhees who was already running quite a large company that still bears his name. It was valuable to see the ideas of transport modelling put to work. I think that experience reinforced my commitment that modelling was a contribution to achieving things, to making use of the science. In addition to Alan himself, I met Walter Hansen – who was working for Voorhees and was probably the inventor of the concept of 'accessibility' in modelling through his paper 'How accessibility shapes land use' – as well as T. R. ('Laksh')

Lakshmanan of the 'Lakshmanan and Hansen' retail modelling paper, another critical and ever-present part of the toolkit.[5] I travelled alone to the Voorhees' Bethesda offices. For some reason this Washington visit was supported by the British Embassy, who arranged for a car and driver to take me out to Bethesda. In the event this was a large Daimler, almost certainly the Ambassador's car. When I left, Alan and his colleagues came to the entrance to see me off and there was great hilarity at the sight of the enormous car!

From a different part of the agenda in Washington, I met Clopper Almon. He was working for the US government (as I remember) on regional input–output models, one of the few economists to be working in this field. My meeting with Almon encouraged me to join this minority in years to come.

Much of what I learned on that trip has remained as part of my intellectual toolkit. Much of it has led to long-standing exchanges – particularly through regional science conferences. Some has led to close working collaboration. Ten years later, Brit and Walter recruited me between them to the position of Adjunct Professor in Regional Science at Penn, where I spent a few weeks every summer in the late 1970s. I worked closely with Brit and those visits were the basis for my work with him on urban dynamics that was published in 1978. It is still a feature of my ongoing work plan. I could chart a whole set of contacts and collaborations for subsequent decades. Maybe the starting points are always influential for any of us, but I was very lucky in one particular respect – it was the start of the modern period of urban modelling and there was everything to play for. I was also very fortunate in being able to meet the key research figures of the time. I suspect that now, as fields have expanded, it would be much more difficult for a young researcher to do this.

Against oblivion: mining the past

I was at school in the 1950s – Queen Elizabeth Grammar School, Darlington – with Ian Hamilton. He went on to Oxford and became a significant and distinguished poet, critic, writer and editor – notable, perhaps, for shunning academia and running his editorial affairs from the Pillar of Hercules public house in Greek Street in Soho. I can probably claim to be the first publisher of his poetry as Editor of the school magazine – poems that, to my knowledge, have never been 'properly' published. We lost touch after school. Ian went on to national service and Oxford; I deferred national service and went to Cambridge. I think we only met once in later years – by coincidence on an underground station platform in the 1960s or 1970s. He died at the end of 2001.

However, I did follow Ian's work over the years and one of his books recently gave me food for thought – *Against Oblivion*, published posthumously in 2002.[6] This book contained brief lives of 50 poets of the twentieth century, emulating a work by Dr Johnson concerning poets of the seventeenth and eighteenth centuries. He also refers to two earlier twentieth-century anthologies. The title of Ian's book reflects the fact that a large proportion of the poets in these earlier selections have now disappeared from view – a point he checked with friends and colleagues – and descended into oblivion. He took this as a warning about what would happen to the reputations of those included in his selection a hundred years into the future and by implication, the difficulty of making a selection at all. It is interesting to speculate about what survives – whether oblivion is in some sense just or unjust. Were those who have disappeared from view simply 'fashionable' at the time – see the section *Following fashion* below (pp. 86–9) – or do they represent a real loss?

This has made me think about 'selection' in my own field of urban modelling. I edited a five-volume 'history' of urban modelling by selecting what I judged – albeit subjectively – to be significant papers and book extracts; these were then published in more or less chronological order.[7] The first two volumes cover around the first 70 years and include 70 or so authors. Looking at the selection again, particularly for these early volumes, I remain reasonably happy with it, though I have no doubt that others would do it differently. Two interesting questions then arise: which of these authors would still be selected in 50 or 100 years' time? Who have we missed and who should be rescued from oblivion?

The first question cannot be answered, only speculated about. It is possible to explore the second, however, by scanning the notes and references at the end of each of the published papers. Such a scan reveals quite a large army of researchers and early contributors. Some of them were doing the donkey work of calculation in the pre-computer age, but many, as now, were doing the 'normal science' of their age. It is this normal science, the constant testing and retesting of ideas, old and new, that ultimately gives fields their credibility. However, I am pretty sure there are also nuggets, some of them gold, to be found by trawling these notes and references. This might be called 'trawling the past for new ideas', but it is a kind of work which is, on the whole, not done.

Such a task would be closely related to delving into, and writing about, the history of fields, which in urban modelling has only be done on a very partial and selective basis, through review papers in the main. (Though the thought occurs to me that a very rich source would be the obligatory literature reviews and associated references in PhD theses. I am not an enthusiast for these reviews as Chapter 1 of theses because they do not usually make for an interesting read, but this argument suggests that they may have tremendous potential value as appendices.) There is one masterly exception – *Forecasting Urban Travel* by Dave Boyce and Huw Williams. While very interesting in fulfilling its prime aim as a history of transport modelling[8], this book would also act as a resource for trawling the past to see what we have missed. This kind of history also involves selection, but when thoroughly accomplished, as in this case, it is far more wide ranging.

Most of us spend most of our time doing normal science. We recognise the breakthroughs and only time will tell whether they survive or are in turn overtaken – whether they were substantive game-changers or not. In his introduction to *Against Oblivion*, Ian Hamilton provides some clues about how this process works – confirming that it is, at least, a process worth studying. For me it suggests a new kind of research – trawling the past for half-worked out ideas that may have been too difficult at the time and could be resurrected and developed.

Following fashion

Choosing a research topic – even a research field – is difficult and tricky. Much research follows the current fashion. This leads me to develop an argument around two questions: how does 'fashion' come about and how should we respond to it?

My research career began in transport modelling in the 1960s. It was driven forward from an academic perspective by my work on entropy-maximising and from a policy perspective by working in a group of economists on cost-benefit analysis. Both modelling and cost-benefit analysis were the height of fashion at the time. I did not choose transport modelling – it was an available job after many failed attempts to find social science employment as a mathematician. The fashionability of both fields was almost certainly rooted in the highway building programme in the United States in the 1950s. There was a need for good investment appraisal of large transport projects. As noted above (pp. 82–4), planners such as T. R. ('Laksh') Lakshmanan and Walter Hansen developed concepts such as accessibility and retail models. This leads me to a first conclusion that fashion can be led from either the academic side or the 'real' policy side – in my case, perhaps unusually, it was from both. Those involved – probably from both sides – realised pretty quickly that transport modelling and land use were intertwined and so, led by people such as Britton Harris and Jack Lowry, the comprehensive urban modelling field was launched. I joined this enthusiastically.

These narrower elements of fashion were matched by a broader social science drive to quantitative research, though probably the bulk of this was statistical rather than mathematical. It is interesting to review the contributions of different disciplines – something that would make a good research topic in itself. The quantitative urban geographers were important: Peter Haggett, Dick Chorley, Brian Berry, Mike Dacey, Les Curry and others – a distinguished and important community. They introduced the beginnings of modelling, but were not themselves modellers.[9] The models grew out of engineering. The economists were surprisingly unquantitative. Walter Isard initiated and led the interdisciplinary movement of 'regional science' which thrives today.[10] From a personal point of view, I moved into geography as a good 'broad church' base. I was well supported by research council grants and built a substantial modelling research team.

By the late 1970s and early 1980s, however, I had become unfashionable which is possibly an indicator of the half-life of fashions. There were two drivers: the academic on the one hand and planning and policy on the other. There was Douglas Lee's 'Requiem for large-scale models'[11] (which seemed to me to be simply anti-science but was influential) and a broader Marxist attack on 'positivist' modellers – notwithstanding the existence of distinguished Marxist modellers such as Sraffa. And model-based quantitative methods in planning – indeed to an extent, planning itself – became unfashionable around the time of the Callaghan government in the late 1970s. Perhaps, and probably, as modellers we had failed to deliver convincingly.

By the mid-1980s, research council funding having dried up, I decided, together with a colleague, Martin Clarke, to explore the prospect of 'going commercial' as a way of replacing the lost research council funding. That story is told in Chapter 6 (pp. 75–7) and it was successful – after a long 'start-up' struggle. As in the early days of modelling, in racing parlance, we had 'first mover' advantage and we were valued by our clients. It would be difficult to reproduce this precisely now because so much of the expertise has been internalised by the big retailers. But that was one response to becoming unfashionable.

By the 2000s, complexity science had become the new fashion. I knew I was a complexity scientist as an enthusiastic follower of Warren Weaver and I happily rebadged myself. This led to new and substantial research council funding. In effect, modelling became fashionable again but under a new label (also supported by the needs of environmental impact assessment in the United States, which needed modelling). By the 2010s the complexity fashion was already fading and new responses became needed. We need now to examine the most recent fashions and consider what they mean for research priorities.

Examples of current fashions are: agent-based modelling (ABM); network analysis; study of social media; big data; and smart cities. The first three are academic-led, the fourth is shared and the fifth is policy and technology-led (unusually by large companies rather than academia or government). The first two have some substantial, interesting ideas, but are on the whole carried out by researchers who have no connection to other styles of modelling. They have not made much impact outside academia. In the ABM case it is possible to show that with appropriate assumptions about 'rules of behaviour' the models are equivalent to more traditional (but under-developed) dynamic models. It may also be the case that, as a modelling technique, ABM is more suited to the finer scale – for example, pedestrian modelling in shopping precincts. ABM is sometimes confused with microsimulation – a field that is developing and deserves to be a new fashion. It also offers scope for major investment.

A curiosity of the network analysis is a focus on topology, to such an extent that valuable and available information is not used. For example, in many instances flows between nodes are known (or can be modelled). They can also be loaded onto networks to generate link loads, but this rich information is not usually used by network analysts. This is probably a failure of the network community to connect to – or even to be aware of – earlier and relevant work. In this case, as in others, there are easy research pickings to be had simply by joining up.

The large-scale study of social media is an interesting phenomenon. I suspect it is done because there are large sources of data that can then be plugged into the set of network analysis techniques mentioned earlier. If this could be seen as modelling telecommunications as an important addition to the comprehensive urban model, it would be valuable both as a piece of analysis and for telecoms companies and regulators. Unfortunately these connections are not typically made. Interestingly flows are not usually modelled and there are research opportunities here (and in the related field of Internet of Things).

The big data field is clearly important, but its significance should be measured against the utility of the data in analysis and policy, rather than as a field in itself. This applies to the growing 'discipline' of data science – if this develops as a silo, the full benefits of new data sources will not be collected. However, there is a real research issue to be discussed here: the design and structure of information systems that connect big data to modelling, planning and policy analysis.

The 'smart cities' field is important in that all efficiency gains are to be welcomed. But it is a fragmented field, mostly focused on very specific applications – even down to the level of 'smart lamp posts' – and there is much thinking to be done in terms of integration with other forms of analysis and planning and being smart for the long run.

There is one general conclusion to be drawn that I will emphasise very strongly: fashion is important because usually (though not always) it is a recognition of something important and new; but the degrees of swing to fashion are too great. There are many earlier threads which form the elements of core social science but have become neglected. Fortunately, there is usually a small but enthusiastic group who keep the important things moving forward, so that the foundation is there for when those threads become important and fashionable again (albeit sometimes under another name). In choosing research topics, it is thus important to be aware of the whole background and not just of what is new. Sometimes integration is possible. Sometimes the old has to be a continuing and developing thread. It is also worth asking when a fashion ceases to be 'new'. The moral may be that if you choose to follow a fashion, get in early and be reasonably confident that it will be fruitful. If you are late, the practising community will already be large and very competitive.

Venturing into other disciplines

Urban and regional science – a discipline or a subdiscipline, or is it still interdisciplinary? – has been good at welcoming people from other disciplines, notably, in recent times, physicists. Can we venture outside our box? It would be rather good if we could make some valuable contributions to physics. However, given that the problems we handle are in some sense generic, such as spatial interaction, we can look for similar problems in other disciplines and see if we can offer a contribution. I can report several of my own experiences which give some clues on how these excursions can come about and may be food for thought for something new for others.

I ventured into demography with Phil Rees many years ago,[12] and I tried to improve the Leontief-Strout inter-regional input–output model around the same time. The latter has been implemented only once by Geoff Hewings and colleagues.[13] Some of this has re-emerged in the *Global Dynamics* project, so it is still alive.[14] Both of these are broadly within regional science. More interesting examples are in ecology, archaeology, history and security. All relate to spatial interaction and competition-for-resources modelling by some combination of 1) adding space, 2) applying the models to new elements, or 3) thinking of new kinds of flow for new problems. In some cases, our own core models have to be combined with those from the other field. But let us be specific.

The oldest example dates back to the mid-1980s; after a gap in time, it has proved one of the most fruitful. Around 1985 Tracey Rihll, then a research student in ancient history in Leeds, came to see me in the Geography Department. She said that someone had told her I had a model that would help her with her data. The data were points representing the locations of known settlement in Greece around 800 BC. What we did was make some colossal assumptions about interactions – say trade and migration – between settlements and then, using Euclidean distance, run the data through a dynamic retail model to estimate – at equilibrium – settlement sizes. Out popped Athens, Thebes, Corinth, etc. – somehow teased out from the topology of the points. One site was predicted as large that had not been thought to be so and if we had had the courage of our convictions we would have urged archaeologists to go there. We published three papers on this work.[15] Nothing then happened for quite a long time until it was picked up, first by some American archaeologists and then by Andy Bevan in UCL Archaeology. Somehow the penny dropped with Andy that the 'Wilson' of 'Rihll and Wilson' was now in UCL and we began to work together – first reproducing the old results and then extending them to Crete.[16] These methods were then separately picked up in UCL by Mark Altaweel in Archaeology and Karen Radner in History; around 2013 we started working on data from the Kurdistan part of Iraq. In this case, the archaeologists really were prepared to dig at what the model predicted were the largest sites. Sadly this has now been overtaken by events in that part of the world. This work has led to further published papers.[17] There is also one other archaeology project, but this links with 'security' below.

The excursion into ecology came about in a different way. The dynamic retail model, as we have noted, is based on equations very similar to the Lotka-Volterra equations in ecology. I thus decided to investigate whether there was the possibility of knowledge transfer between the two fields. What was most striking to me in ecology was that virtually all the applications were aspatial, notwithstanding the movement of animals and seeds as a kind of spatial interaction. I was therefore able to articulate what a spatial L-V system might look like in ecology. I did not have the courage to try to publish it in an ecology journal, but *Environment and Planning A* published the paper, which I fear has fallen rather flat.[18] Nevertheless I still believe that it is important. Indeed, it is highly relevant in exploring the spatial dynamics of the COVID pandemic.[19]

There was a different take on history with the work I did with Joel Dearden[20] [21] on Chicago. This came about because I had been invited to give a paper at a seminar in Leeds to mark Phil Rees's retirement and I had been reading William Cronon's book, *Nature's Metropolis*, about the growth of Chicago.[22] Cronon's book charted in detail the growth of the railway system in North America in the nineteenth century. Phil had done his PhD in Chicago and so this seemed like a good topic for the seminar. Joel and I designed a model – not unlike the Greek one, but in this case with an emphasis on the changing accessibility provided by the growth of railways over the century. We had US Census data from 1790 against which we could do some kind of checking and we generated a plausible dynamics. This work is being taken forward in the UK context as part of the *Living with machines* project in The Alan Turing Institute.

The fourth area was 'security', stimulated by this being one of the four elements of the EPSRC *Global dynamics* project that I was working on at the time. This turns on the interesting idea of interpreting spatial interaction, as 'threat' that can attenuate with distance. From a theoretical point of view, the argument was analogous to the ecological one. Lewis Fry Richardson, essentially a meteorologist, developed an interest in war in the 1930s and built models of arms races, in effect using L-V models.[23] Again, however, this took place without any spatial structure. We have now been able to add space, making the model much more versatile, and we have applied it in a variety of situations.[24] We even reconnected with archaeology and history again by seeking to model, in terms of threat, the summer 'tour' of the Emperor of Assyria, with his army, in the Middle Bronze Age, taking over smaller states and reinforcing existing components of the Empire. In this case, the itineraries of the tour were recorded on stone tablets – an unusual source of data for modellers.[25] The models are currently being applied in a contemporary context through a project in The Alan Turing Institute.[26]

All of these ventures have been modest. Two of them were funded by small UCL grants and as small parts of larger projects in The Alan Turing Institute. Papers have been accepted and published in relation to all of them. However, attempts to obtain funding from research councils have mostly failed. The Turing project is an exception, but that is part of something larger. Are we ahead of our time? Or (possibly more likely) are the community of available referees unable to handle this kind of interdisciplinarity, particularly if algebra and calculus are involved?

What would Warren Weaver say now? The next game-changers

Warren Weaver was a remarkable man. A distinguished mathematician and statistician, he made important early contributions on the possibility of the machine translation of languages. He was also a fine writer who recognised the importance of Shannon's work on communications and the measurement of information, and he worked with Shannon to co-author *The Mathematical Theory of Communication*.[27] Perhaps above all, however, he was a highly significant science administrator. For almost 30 years, from 1932, Weaver worked in senior positions for the Rockefeller Foundation, latterly as Vice-President. I guess he had quite a lot of money to spend. From his earliest days with the Foundation, he evolved a strategy which was potentially a game-changer, or at the very least seriously prescient – he switched his funding priorities from the physical sciences to the biological. In 1948 he published a famous paper in *The American Scientist* that justified this – maybe with an element of post hoc rationalisation – based on three types of problem (or three types of system, according to taste): simple, of disorganised complexity and of organised complexity.

Simple systems have a relatively small number of entities; complex systems have a very large number. The entities in the systems of disorganised complexity interact only weakly; those of organised complexity have entities that interact strongly. In the broadest terms – my language not his – Newton had solved the problems of simple systems and Boltzmann those of disorganised complexity. The biggest research challenges, he argued, were those of systems of organised complexity; more of these were to be found in the biological sciences than the physical.[28] How right he was! It has only been after some decades that 'complexity science' has come of age – and become fashionable. As noted earlier, I was happy to re-badge myself as a complexity scientist, which may have helped me to secure a rather large research grant.

There is a famous management scientist, no longer alive, called Peter Drucker. Such was his fame that a book was published confronting various business challenges with the title *What Would Peter Drucker Say Now?* Since no one, to my knowledge, has updated Warren Weaver's analysis, I am tempted therefore to pose the question 'what would Warren Weaver say now?' I have used his analysis for some years to argue for more research on urban dynamics – recognising cities as systems of organised complexity. But let us explore the harder question. Given that we understand urban organised complexity, though we have not progressed a huge distance with the research challenge, if Warren Weaver was alive now and could invest in research on cities, could we imagine what he might say to us? What might the next game-changer be? I will argue it for 'cities' but, *mutatis mutandis*, the argument could be developed for other fields. Let us start by exploring where we stand against the original Weaver argument.

We can probably say a lot about the 'simple' dimension. Computer visualisation, for example, can generate detailed maps on demand which can provide excellent overviews of urban challenges. We have done pretty well on the systems of disorganised complexity in areas like transport, retail and beyond. This has been done not only in an explicit, Boltzmann-like way with entropy-maximising models, but also with various alternatives from random utility models via microsimulation to agent-based modelling (ABM). We have made a start on understanding the slow dynamics with a variety of differential and difference equations, some with roots in the Lotka-Volterra models, others connected to Turing's model of morphogenesis. What kinds of marks would Weaver give us? Pretty good on the first two: making good use of dramatically increased computing power and associated software development. I think on the disorganised complexity systems, when he saw that we have competing models for representing the same system, he would tell us to get that sorted out and either decide which is best and/or work out the extent to which they are equivalent or not at some underlying level. He might add one big caveat – we have not applied this science systematically and we have missed opportunities to use it to help tackle major urban challenges. On urban dynamics and organised complexity we would probably get marks for making a goodish start, but with a recommendation to invest a lot more.[29]

So we still have a lot to do, but where do we look for the game-changers? Serious application of the science – equilibrium and dynamics – to the major urban challenges could be a potential game-changer. A full development of the dynamics would open up the possibility of 'genetic planning' by analogy with 'genetic medicine'. For the new, however, I think we have to look to rapidly evolving technology. I would single out two examples, already introduced in Chapter 5, and there may be many more. The first is in one sense already old hat: big data. However, I want to argue that if it can be combined with high-speed analytics, this could be a game-changer. The second is something which is entirely new to me and may not be well known in the urban sciences: block chains. A block is some kind of set of accounts at a node. A block chain is made up of linked nodes – a network. There is much more to it, however, and it is being presented as a disruptive technology that will transform the financial world. A block chain transfers money. Could it transfer data and solve personal data security challenges at the same time? If you google the term, you will find out that it is almost wholly illustrated by the bitcoin system. A challenge is to work out how it could transform urban analytics and planning.

However, the big question is worth further exploration: what would Warren Weaver say now? This leads me to something I have argued for a long time without any real success: treat the social sciences as 'big science'. Weaver switched his funding (simplifying slightly) from physics to biology. Perhaps the major funders – governments and charities – should switch some funding from the 'hard' sciences to the social sciences?

Exercises

1   For your research project, review the history of your topic, as for a literature review but focusing on the game-changers. Where might the future game-changing breakthroughs lie?
2   More broadly, can you identify past researchers who might have left contributions which could now be seen to be valuable but were neglected at the time? (This may have happened because computing power was unavailable, for example.)
3   How is the literature in your field dominated by some current fashion? Or not?
4   Can you see any ways in which the methods or results of your current project could be applied in another discipline?

Notes

1   See Foster and Beesley, 'Estimating the social benefit of constructing an underground railway in London'.
2   Isard, *Location and the Space-Economy*.
3   Herbert and Stevens, 'A model for the distribution of residential activity in an urban area'.
4   Senior and Wilson, 'Explorations and syntheses of linear programming and spatial interaction models of residential location'.
5   Lakshmanan and Hansen, 'A retail market potential model'.
6   Hamilton, *Against Oblivion*.
7   Wilson, ed., *Urban Modelling: Critical concepts in urban studies*.
8   Boyce and Williams, *Forecasting Urban Travel*.
9   Haggett, *Locational Analysis in Human Geography* was perhaps the iconic quantitative geography book of the period.
10   Isard, *Methods of Regional Analysis*.
11   Lee, 'Requiem for large-scale models'.
12   Rees and Wilson, *Spatial Population Analysis*.

13 Kim, Boyce and Hewings, 'Combined input–output and commodity flow models for interregional develoment planning'.
14 Wilson (ed.), *Global Dynamics; Approaches to Geo-Mathematical Modelling*.
15 Rihill and Wilson, 'Spatial interaction and structural models in historical analysis'; 'Model-based approaches to the analysis of regional settlement structures'; 'Settlement structures in Ancient Greece'.
16 Bevan and Wilson, 'Models of settlement hierarchy based on partial evidence'.
17 See, for example, Davies et al., 'Application of an Entropy Maximizing and Dynamics Model for Understanding Settlement Structure'.
18 Wilson, 'Ecological and urban systems models'.
19 Wilson, 'Epidemic models with geography'.
20 Sadly Joel died in 2020.
21 Wilson and Dearden, 'Tracking the evolution of regional DNA'.
22 Cronon, W., *Nature's Metropolis*.
23 Richardson, *Arms and Insecurity*.
24 Baudains and Wilson, 'Conflict modelling'.
25 Baudains, P., Zamazalová, S., Altaweel, M. and Wilson, A. 'Modeling strategic decisions in the formation of the Early Neo-Assyrian Empire'.
26 Guo, W., Gleditsch, K. and Wilson, A., 'Retool AI to forecast and limit wars', *Nature*, 562, 331–33.
27 Shannon and Weaver, *The Mathematical Theory of Communication*.
28 Weaver, 'Science and complexity'; 'A quarter century in the natural sciences'.
29 Investment in urban research is small beer compared, for example, to Google's investment in Deep Mind.

Chapter 8
Adding to the toolkit 2: more on superconcepts

Introduction

Implicitly, over the years, it is possible to build an intellectual toolkit – or, in the research context, a research toolkit – which provides the concepts that can be brought to bear on a range of challenges as they emerge. Important elements of that toolkit are the superconcepts, introduced in Chapter 2, which cross disciplines and can be applied to a range of what turn out to be generic problems. We have argued that most of the big research problems are interdisciplinary and the toolkit kit can consequently become very important – continually expanding, of course. The importance here is because it draws on the *breadth* of knowledge required as well as the depth – one of the core challenges of working in an interdisciplinary way.

The core elements of my own toolkit (STM and PDA, for example) have been sketched above, particularly in Chapters 1 and 3. What follow are ideas that have been added as subsidiary elements, drawing from the superconcepts toolkit, adding to the introduction of these in Chapter 2. These examples are personal, of course, but readers are urged to develop their own. I discuss in turn:

- The brain as a model (pp. 97–8).
- 'DNA' (pp. 99–100).

- Territories and flows (pp. 100–2).
- Optimisation (pp. 102–4).
- Missing data (pp. 104–6).

These are all relevant to a wide range of research challenges and are also interesting in their own right.

The brain as a model

I will start with Stafford Beer's book *Brain of the Firm*, which has ideas I have used since it was first published in 1972.[1] A larger-than-life character, he was a major figure in operational research, cybernetics, general systems theory and management science. I have a soft spot for him because, although I never met him, he wrote to me in 1970 after the publication of my *Entropy* book saying that it contained the best description of entropy he had ever read. His 'brain' book is still in print as a second edition; I think it is also possible to download a pdf. Googling will also fill in more detail on Stafford Beer – but beware the entry to the 'Stafford Beer Festival 2015'. This first made me think he still had a contemporary supporters' club, until I discovered it was actually for the *Stafford* Beer and Cider Festival.

The core argument of the book is a simple one: that the brain is the most successful system ever to evolve in nature and therefore, if we explore it, we might learn something. In *Brain*, Beer expounds the neurophysiology to a point when, at the time of first reading, it seemed so good that I checked the accuracy against some neurophysiology texts. It seemed to pass. What follows is a considerable oversimplification, both of the physiology and of Beer's use of it. The brain has five levels of organisation. The top, Level 5, is the strategic level. Levels 1–3 represent the autonomous nervous system which govern actions such as breathing without us having to think about it; the system also carries instructions to perform actions determined at Level 1. Occasionally the autonomous system passes messages upwards – if there is some danger, for example. Level 4 is particularly interesting. It can be seen as an information processor. The brain receives an enormous amount of data which it could not make sense of without this filter. Beer's argument is that there is no equivalent function in organisations – to their fundamental detriment. He cites as an exception to this the Cabinet War Rooms in the UK in the Second World War (now open as part of the Imperial War Museum).

The War Rooms were specifically set up to handle the real-time flow of information and to deal with information overload.

Beer translated this into a model of an organisation which he called the VSM – the viable system model.[2] The workings of the organisation took place at Levels 1, 2 and 3. The top level – the company board or equivalent – was Level 5. He usually attributes Level 4 to the Development Directorate and I can see the case for that, but it does not entirely deal with the filtering operation that any organisation needs. (This is probably because of an over-rapid re-reading on my part.) However, what he did recommend, even in 1972, was 'a large dynamic electrical display of the organisation', together with a requirement that all meetings of senior staff took place in that room. The technical feasibility of this is now much higher and fits with the display of big data, such as that built in Glasgow as an Innovate UK demonstrator.

This still leaves open the question of how to make sense of the mass of data post-filtering, and this is where we need an appropriate model. This connects to a little explored research question: how to design the architecture of a multi-dimensional information system that can be aggregated and interrogated in a variety of ways.

I think we can gain tremendous insights from the *Brain of the Firm* model when we think about organisations, either those we work in or for or those we are simply interested in. Additionally, can we learn anything about how to approach research? We are certainly aware of information overload and functioning as individual researchers, the scale of which can makes it impossible to cope. Can we build an equivalent of a War Room? Forward-looking librarians are probably trying to help us by doing this electronically – but we run into the classification problem. As a student I attended a Cambridge Moral Sciences (aka 'philosophy') Club seminar in which Professor Margaret Masterman observed that the library classification problem was of the same order as that of the machine translation of languages. Decades later I repeated her comment in a British Academy seminar. Karen Spärck Jones, a student at the time of the seminar, said, 'I was there,' and sent me some papers on the subject.

Can we organise any other kind of cooperative effort – a form of crowd sourcing to find the game-changers out of the mass of information on research? We might call this the 'market in research'. The buyers are the researchers who cite other research and a cumulatively large number of citations usually points to something important. The only problem then is that the information comes too late. We learn late about the new fashion; we do not get in on the ground floor. There is an unresolved challenge here.

'DNA'

The idea of 'DNA' has become a commonplace metaphor. The real DNA is the genetic code that underpins the development of organisms. I find the idea useful in thinking about the development – or evolution – of cities. This development depends very obviously on 'what is there already' – in technical terms we can think of that as the 'initial conditions' for the next time step in a dynamic model. We can then make an important distinction between what can change quickly, such as the pattern of a journey to work, for instance, and what changes only slowly such as the pattern of buildings or a road network. It is the underpinnings of the slowly changing stuff that represents urban DNA. Again, in technical terms, it is the difference between the fast dynamics and the slow dynamics. The distinction is between the underlying *structure* and the *activities* that can be carried out on that structure.

It also connects to the complexity science picture of urban evolution, particularly the idea of path dependence. How a system evolves depends on the initial conditions. Path dependence is a series of initial conditions. We can then add that if there are nonlinear relations involved – economies of scale, for example – then the theory shows us the likelihood of phase changes – abrupt changes in structure. The evolution of supermarkets is one example of this; gentrification is another.

This offers another insight – future development is *constrained by the initial conditions*. We can therefore ask the question of what futures are possible – given plans and investment – from a given starting point? This is particularly important if we want to steer the system of interest towards a desirable outcome or away from an undesirable one – also a tricky challenge here, taking account of possible phase changes. This then raises the possibility that if we can change the DNA, we can invest in such a way as to generate new development paths. This would be the planning equivalent of genetic medicine – a concept which could be seen as 'genetic planning'.

There is a related and important discovery from examining retail dynamics from this perspective. Suppose there is a planned investment in a new retail centre at a particular location, for example: this constitutes an addition to the DNA. The dynamics then show that this investment has to exceed a certain critical size for it to succeed. If this calculation could be done for real-life examples (as distinct from proof-of-concept research explorations), this would be incredibly valuable in planning contexts.[3] Intuition suggests that a real-life example might be the initial investment in Canary Wharf in London. In the end that proved big enough to pull with it a tremendous amount of further investment. The same thing may be happening with the Crossrail investment in London, around stations such as Farringdon.

The 'structure vs activities' distinction may be important in other contexts as well. It has always seemed to me that it is worth distinguishing in a management context between 'maintenance' and 'development', and keeping these separate. There is thus a demarcation between keeping the organisation running as it is and planning the investment that will shape its future.

The DNA idea can be part of our intuitive intellectual toolkit, and can then function more formally and technically in dynamic modelling. The core insight is worth having.

Territories and flows

Territories are defined by boundaries at scales ranging from countries (and indeed alliances of countries) to neighbourhoods via regions and cities. These may be government or administrative boundaries, some formal, others less so, or they may be socially defined, for example in gang territories in cities. Much data relates to territories and some policies are defined by them – catchment areas of schools or health facilities, for example.

It is at this point that we start to see difficulties. Local government boundaries usually will not coincide with the functional city region; in the case of catchment boundaries, some will be crossed unless there is some administrative 'forcing'. So as well as defining territories, we need to consider flows both within – but especially between – them. Formally, we can call territories 'zones', and flows are then between origin zones and destination zones. If the zones are countries, then the flows are trade and migration; if zones within a city region, then the flows may be journeys to work, to retail or other facilities.

It is thus convenient to make a distinction between the social and political roles of territories and how we can best make use of them in analysis and research. In the former case, much administration is rooted in the government areas and they have significant roles in social identity – 'I am a Yorkshireman or woman', 'I am Italian', and so on. In the latter case, these territories do not typically suit our purposes, though we are often prisoners of administrative data and associated classifications.

So how do we make the best of it for our analysis? A part of the answer is always to make use of the flow data. In the case of functional city regions, the whole region can be divided into smaller zones and origin-destination flows (O-D matrices technically) can be analysed, first to identify 'centres' and then perhaps to highlight a hierarchy of centres[4] for one way to do this systematically. It then becomes possible, for example, to define a city region as a 'travel to work area' – a TTWA – as in the UK census. Note that there will always be an element of arbitrariness in this, however. What is the cut-off – the percentage of flows from an origin zone into a centre – that determines whether that origin is in a particular TTWA or not?

In analysis terms, I would argue that the use of flow data is always critical. Very few territories – zones at any scale – are self-contained. And the flows across territorial boundaries, as well as the richer sets of O-D flows, are often very interesting. An obvious example is imports and exports across a national boundary from which the 'balance of payments' can be calculated – revealing something about the health of an economy. In this case, the data exists (for many countries), but in the case of cities, it does not. Yet the balance of payments for a city (however defined) is a critical measure of economic health. There is a big research challenge there.

It is helpful to point to some contrasts in both administration and analysis when flows are not taken into account and then to consider what can be done about this. There are many instances when catchments are defined as areas inside fixed boundaries, even when they are not defined by government. Companies, for example, might have CMAs (customer market areas) whilst primary schools might draw a catchment boundary on a map giving priority to 'nearness' but trying to ensure that they get the correct number of pupils. In some traditional urban and regional analysis – in the still influential Christaller central place theory, for example – market areas are defined around centres; in Christaller's case nested in a hierarchy. This makes intuitive sense, but has no analytical precision because the market areas are not self-contained. As it happens, there is a solution.[5]

Think of a map of facilities – shopping centres, hospitals, schools or whatever. For each, add to the map a 'star' of the origins of users, with each link being given a width to represent that number of users. For each facility, that star is the catchment population. It all adds up properly – the sum of all the catchment populations equals the population of the region. This, of course, represents the situation as it actually is, which is fine for retail analysis, for example. It is also fine for the analysis of the location of health facilities. It may be less good for primary schools that are seeking to define an admissions policy.

A particular application of the 'catchment population' concept is in the calculation of performance indicators. If cost of delivery per capita is an important indicator, this can then be calculated as the cost of running the facility divided by the catchment population. It is clearly vital that there is a good measure of catchment population. In this case the 'star' is better than the 'territory'. However, the concept can be applied the other way round. Focus on the population of a small zone within a city and then build a reverse star and link to the facilities serving that zone, each link weighted by what is delivered. What you then have is a measure of effective delivery. By dividing by the zonal population, you have a per capita measure. (An alternative, and related, measure is 'accessibility'.) This may sound unimportant, but consider, say, supermarkets and dentists. On a catchment population basis, any one of these facilities may be performing well. On a delivery basis to a population, analysis will turn up areas that are 'supermarket deserts' (usually where poorer people live – those who would like access to the cheaper food) or that have poor access to dental treatment even though the facilities themselves are perfectly efficient.[6]

So what do we learn from this? It is clear that we have to work with territories, because they are administratively important and may provide the most data, but we should always, where at all possible, make use of all the associated flows, many of which cross territorial boundaries. We can then calculate useable concepts such as catchment populations and delivery indicators 'properly'.

Beware of 'optimisation'

The idea of 'optimisation' is basic to lots of things we do and to how we think. When driving from A to B, for example, what is the optimum route? When we learn calculus for the first time, we quickly come to grips with the maximisation and minimisation of functions. This is professionalised within operational research. If you own a transport business, you have to plan a daily schedule of collections and deliveries. How do you allocate your fleet to minimise costs and hence to maximise profits for the day? In this case, the mathematics and the associated computer programs exist and are well known – they will solve the problem for you. You have the information and you can control what happens. Now suppose that you are an economist and want to describe and model human behaviour. Suppose you want to investigate the economics of the journey to work. This is another kind of scheduling problem, except that in this case it involves a large number of individual decision makers. If we turn to the micro-economics textbooks, we find the answer: define a utility

of administrative data and much survey data. The real-time data – for example, positional data from mobile phones – can be used to estimate person flows, taking over from expensive (and infrequent) surveys. In practice, of course, much of the data available to us is sample data and we can use statistics – either directly or to calibrate models – to complete the set.

My own early work in urban modelling was a kind of inversion of the missing data problem. Entropy-maximising generated a model which provided the best fit to what is known – in effect a model for the missing data. It turns out, not surprisingly, to have a close relationship to Bayesian methods of adding knowledge to refine beliefs. In theory this only works with large 'populations', but there have been hints that it can work quite well with small numbers. This only gets us so far, however. The collection (or identification) of data to help us build dynamic models is more difficult. Even more difficult is connecting models that rely on averaging in large populations with those that use more micro-scale 'data' – perhaps qualitative information – on individual behaviour. There are research challenges to be met here.

There are other kinds of challenges too, such as what to do when critical elements of a simpler nature are missing. An example is the need for data on 'import and export' flows across urban boundaries, to be used in building input-output models at the city (or region) scale. We need these models so that we can work out the urban equivalent of the well-understood 'balance of payments' in the national accounts. How can we estimate something which is not measured at all, even on a sample basis? I recently started to ponder whether we could look at the sectors of an urban economy and make some bold assumption that the import and export propensities were identical to the national ones. This immediately throws up another problem, however; we have to distinguish between intra-national – that is, between cities – and international flows. It became apparent pretty quickly that we needed the model framework of interacting input-output models for the UK urban system before we could progress to making estimates – albeit very bold ones – of the missing data. We have done this for 200+ countries in a global dynamics research project.[13] The task now was to translate this to the urban scale, but for the country as a whole. A 'missing data' problem may be seen to be quite a tricky theoretical one.

Perhaps the best way to summarise the 'missing data' challenges is to refer back to the requisite knowledge argument of Chapter 1 – what is the 'requisite data set' needed for an effective piece of research, for example to calibrate a model? If the model is good, the model outputs look like 'data' for other purposes. In more general terms, the message is not to be put off from doing something important because of 'missing data'. There are ways and means around this, albeit sometimes difficult ones.

Exercises

1   Identify the information flows in your system of interest. How can you structure them in an information system for use a) by the analytics researcher and b) by a policy or decision maker who is interested in real challenges? The 'information' may include model outputs as well as data.

2   What are the analogies of DNA in your system of interest? Show how, as 'initial conditions', they partly determine the possibilities of system evolution.

3   Review the spatial structure of your system of interest and identify the territories and flows that will be a fundamental part of the basis of your system description.

4   Is there any element of the theory of how your system works that can be addressed through an optimisation hypothesis? Explore the ways in which this can be handled.

5   How are you going to deal with the fact that some of the data you would like to have is 'missing'?

Notes

1 Beer, *Brain of the Firm*.
2 Beer, *Designing Freedom*.
3 Dearden and Wilson, *Explorations in Urban and Regional Dynamics*.
4 Nystuen and Dacey, 'A graph theory interpretation of nodal regions'.
5 Wilson, Spatial interaction and settlement structure'.
6 Clarke and Wilson, 'Performance indicators and model-based planning I'; 'Performance indicators and model-based planning II'.
7 First introduced in H. A. Simon (1956) Rational choice and the structure of the environment, *Psychological Review*, 63, 129–138: "Evidently, organisms adapt well enough to 'satisfice', they do not in general 'optimize'. p 136.
8 Dearden and Wilson, 'The relationship of dynamic entropy maximising and agent-based approaches'; *Explorations in Urban and Regional Dynamics*.
9 Evans, 'A relationship between the gravity model for trip distribution and the transportation model of linear programming'.
10 See Dennett and Wilson, 'A multi-level spatial interaction modelling framework for estimating interregional migration in Europe', for examples of the use of biproportional fitting in this context.
11 *Nullius in verba* is the Royal Society's motto. It roughly translates as 'take nobody's word for it', emphasising the integrity of data and experiment.
12 I wanted to add something on 'theory' and so have suggested *Evolvere theoriae et intellectum*. This is translated by Google as 'to develop the theory and understanding' – rather more of a mouthful.
13 Wilson, *Global Dynamics*; *Approaches to Geo-Mathematical Modelling*.

Part 4
Managing and organising research

Chapter 9
Managing research, managing ourselves

Introduction

Management is usually thought to be about managing organisations, and of course it is. But it is also, at a micro scale, about managing ourselves as individuals and the same principles apply. In the context of universities, it is in another sense about 'managing ourselves' as the structures are typically dominated by academics.[1] In the following sections I explore various dimensions of management, starting with the observation that most of us learn management from experience, using myself as an example (pp. 110–14). How to encourage 'collaboration' does not normally feature in the standard textbook, but it is particularly important in research (pp. 114–16). Managers also face challenges in allocating resources and I use the competition between 'pure' and 'applied' communities in research to illustrate this (pp. 117–18). Given the power of the big players – the big teams – at both the university or institute scale, or in relation to very well-funded research groups, what can you do to succeed if you are a manager in a small institution or group? Here I take the Leicester City example (pp. 118–19), which demands continually thinking on how to be ahead of the game. How do we know that we have at least achieved 'best practice' (pp. 120–1)? And finally, at the micro-scale, how can we manage our time most effectively and be good writers (pp. 123–5)?

Do you really need an MBA?

'How to manage' has itself become big business as evidenced by the success of university business schools and the MBA badge that they offer (note also the size of the management section of a bookshop). I do not dispute the value of management education or of much of the literature, but my own experience has been different, its primary focus learning on the job. It is interesting, for myself at least, to trace the evolution of my own knowledge of 'management', although I do not claim this to be a recipe for success. There were failures along the way and I have no doubt that there are many alternative routes. I did discover some principles that have served me well, however – most of them filleted from the literature, tempered with experience. This can be thought of as 'research on research', at least at an elementary level.

In my first 10 years of work, my jobs were well focused, with more or less single objectives. 'Management' consisted primarily of getting the job done. This began at the then newly-founded Rutherford Laboratory at Harwell with writing computer programs to identify bubble chamber events at the CERN synchrotron, then moved on to implementing transport models for planning purposes at the then Ministry of Transport (MoT). I set up the Mathematical Advisory Unit in the Ministry, which became large; doing this was clearly a management job. I then moved to the Centre for Environmental Studies (CES) as Assistant Director – another management job. These roles were on the back of a spell of research into urban economics and modelling in Oxford, which also had me working on a broader horizon even when I was a civil servant – CES was, of course, a research institute. From 1964–7 I was an Oxford City Councillor, a Labour member – then a wholly 'part-time' occupation – on top of my other jobs.

What did I learn in those 10 years? At the Rutherford Lab I discovered the value of teamwork and being lightly managed by the two layers above me, as well as that of being given huge responsibilities in the team at a very young age. In MoT I learned something about working in the Civil Service, though my job was well defined; again, I had sympathetic managers above me. On Oxford City Council I learned at first-hand the workings of local government and saw how a political party functioned. This was teamwork of a different kind, whether in council business as a political group or organising to win elections. There may be transferable skills here. At CES I built my own team. At both MoT and CES I recruited some very good people who went on to have distinguished careers. In all cases, the working atmosphere was pretty good.

It was in CES, however, that I realise with hindsight that I made a big mistake. I assumed that urban modelling was the key to the future development of planning and probably convinced Henry Chilver, the director, of the same. But in so doing I neglected the wider context and people such as Peter Wilmott, from whom I should have learned much more. That led to the director standing down as the trustees sought to widen the brief – and to bring it back to its original purpose – and it nearly cost me my job (as I learned from one of the trustees). Fortunately, in the end it did not. The CES was also my first experience of a governing body – the Board of Trustees. Another mistake that I made was to leave the relationship with them entirely with the director, so I had no direct sense of what they were thinking. I left a few months later for the University of Leeds. Within a couple of years I began to have the experience of broader-based management jobs.

I went to Leeds as Professor of Urban and Regional Geography into a school of geography that was being rebuilt through what would then be described as strong leadership, but amounted to what now might be thought of as bullying at times. However, I was left to get on with my job. I enjoyed teaching and was successful in bringing in research grants and I also built up a good team. However, the atmosphere was such that after two or three years I was thinking of leaving. Then, out of the blue, the Head of Department moved on and I found myself in <u>that</u> role.

My first management task was to change the atmosphere. An important element of that was to make the monthly staff meeting really count. The fact that 'leadership' should be through the consent of the staff was a key lesson for me, and something that I could more or less maintain in later jobs. This is not as simple as it sounds, of course; there are often difficult disagreements to be resolved. The allocation of work round the staff of the department was very uneven, for example, and I managed to sort that out with a 'points' system. I learned the value of PR as the first Research Assessment Exercise approached (in 1987) by making sure that we publicised our research achievements. I still believe (without hard evidence of course) that this helped us to a top rating (which was not common in the university at the time).

The geography staff meeting was my first experience of chairing something and I probably learned from my predecessor how not to do it as well as how to ensure that you secure the confidence, as far as possible, of people in the room. I was elected as Head of Department for three three-year spells, alternating with my fellow professor. I began to take on university roles, and in particular the chairing of one of the main committees – the Research Degrees Committee. This in turn led to my becoming Chair of the Joint Board of the Faculties of Arts, Social and Economic Studies and Law – the equivalent of a dean in modern parlance. The Joint Board represented a large chunk of the university and was responsible for a wide range of policy and administration. There were many subcommittees to chair, too, which gave me lots of practice. The board itself had something in the order of 100 members.

In the late 1980s, I was 'elected' – quotation marks are appropriate as there was only one candidate – as Pro-Vice-Chancellor for 1989–91. There was only one such post at the time, and the then VC became Chair of CVCP[2] for those two years, delegating a large chunk of his job to me. The university was not in awful shape, but it was not good either. Every year there was a big argument about cuts to departmental budgets. I began to consider how to turn the 'ship' around – a new management challenge on a big scale. It helped that at the end of my first year as PVC I was appointed as VC-elect from October 1991, so in my second year I could not only plan, but also start to implement some key strategies. This is a long story so I will simply summarise some key elements – challenges and the beginnings of solutions.

The university had 96 departments and was run through a system of over 100 committees (as listed in the university calendar) – a seriously sclerotic structure. For example, there were seven biology departments, each mostly doing molecular biology. We had to shift the climate from one of cost-cutting to one of income generation. This was achieved by delegating budgets to each of a reduced number of departments (96 to 55), a strategy was based on cost management but, critically, with delegated income-generating rules. This went on to become the engine for both growth and transformation. There were winners and losers, of course, which led to some serious handling challenges at the margins. (I tried to resolve these by going to department staff meetings to tackle concerns head on. That sometimes worked, but sometimes did not.) Another challenge was how to marry department plans with the university's own plan. The number of committees was substantially reduced, with a focus on three key committees.

In my first two years as VC, there was a lot of opposition to change, to the point where I even started thinking about the minimum number of years I would have to stay in order to leave in a respectable way. By the third year, many of those who had been objecting had taken early retirement and the management responsibilities around the university – Heads of Department, members of key committees – were being filled by a new generation. I ended up staying in the post for 13 years.

Can I summarise at least some of the key principles I learned in that time?

- Recognising that the university was not a business but had to be business-like
- Maintaining our own strategy within a volatile financial and policy environment; then operating tactically to bring in the resources that we needed to implement this strategy
- Being true to the idea that underpins my thinking about strategy, which I learned from my friend Britton Harris of the University of Pennsylvania in the 1970s – outlined in a different context in Chapter 1. He was a city planner (and urban model builder) and he argued that planning involved three kinds of thinking: policy, design and analysis. He went on to observe that 'you very rarely find all three in the same room at the same time' (p. 6). We can apply this to universities: 'analysis' means understanding the business model and having all relevant information to hand; 'policy' means specifying objectives; and 'design' means inventing possible plans and working towards identifying the best – which then becomes the core strategy. This may well be the most valuable part of my toolkit.
- Recognising that a large institution – by the end of my period 33,000 students and 7,000 staff – could not be run from the centre. An effective system of real delegation was critical.
- Appreciating the importance of informal meetings and discussions held outside the formal committee system. I had meetings at least termly with the main groups: Heads of Department, members of the council, the main staff unions and the Students' Union Executive.
- Committing to openness, particularly of the accounts.
- Acknowledging accountability – in reorganising the committee system, I retained a very large senate, about 200, of whom around half would regularly attend meetings.
- Finally something not in the job description – realising that how I behaved somehow had an impact on the ethos of the university.[3]

By many measures I was successful as a manager and I learned most of the craft as Vice-Chancellor. However, I was constantly conscious of what I was not succeeding at, so I am sure my 'blueprint' is a partial one. I learned a lot from management literature. Mintzberg[4] showed me that if in an organisation your front-line workers are high-class professionals, you would have problems if they did not feel involved in the management. In this respect, I think the university system in the UK has done pretty well but the health service less so. Ashby[5] taught me the necessity of devolving responsibility, Christensen[6] showed me the challenges of disruption and how to work around them. I learned a lot about developing strategy and the challenges of implementation, realising that 'strategy is 5% of the problem, implementation is 95%'.[7] I also learned a lot about marketing. I tried to encapsulate much of this in running seminars for my academic colleagues and for the university administration. Much later I wrote a lot of it up in my book of 2010, *Knowledge Power*.

Bearing all the above in mind, do you really need an MBA? I admire the best of them and their encapsulated knowledge. In my case, I guess I had the apprenticeship version, together with a wide span of management literature and some advice from wise heads.[8] Over time it is possible to build an intellectual management toolkit in which you have confidence that it more or less works. I have tried to stick to these principles in subsequent jobs – UCL, AHRC, The Ada Lovelace Institute, the Government Office for Science, the Home Office and The Alan Turing Institute. Circumstances are always different, however, and the toolkit needs to evolve.

Collaboration

I left the Centre for Advanced Spatial Analysis, part of the Bartlett School of Planning in UCL, in July 2016 and moved to the new Alan Turing Institute. I had planned the move to give me a new research environment – as a Fellow with some responsibility for developing an 'urban' programme. There were few employees. Most of the researchers – part-time academics as Fellows, some Research Fellows and PhD students – were due in October. I ran a workshop on 'urban priorities' and wondered what to do myself with no supporting resources. I was aware that my own research was on the fringes of Turing priorities – 'data science'. I could claim to be a data scientist and indeed Anthony Finkelstein,[9] then a Trustee and a UCL colleague, had said in a conversation encouraging me to move to Turing, 'You can't have "big data" without big models'. However, in Turing, data science meant machine learning and AI rather than modelling as I practised it.

So I started to think about a new niche. After all, Darwin had decided, in his later years, to work on 'smaller problems', perhaps more manageable. I am certainly not comparing myself to Darwin, but there may be good advice there. And as for machine learning, though I put myself on a steep learning curve to learn something new and to fit in, I could not see how I could manage the '10,000 hours challenge' that would turn me into a credible researcher in that field.[10]

At the end of September 2016, everything changed. In odd circumstances, I found myself as the Institute CEO. There was suddenly a huge workload. I reported to a board of trustees, there were committees to work with, there were five partner universities to be visited. Above all, a new strategy had to be put in place – hewed out of a mass of ideas and forcefully-stated disagreements. I can now begin to record what I learned about a new field of research (for me) and the challenges of setting up a new institute. I had to learn enough about data science and AI to be able to give presentations about the Institute and its priorities to a wide variety of audiences. I was able to attend seminars and workshops and talk to a great variety of people and, by a process of osmosis, I began to make progress. I will start by recording some of my own experiences of collaboration in the Institute.

The ideal of collaboration is crucial for a national institute. Researchers from different universities, from industry, from the public sector meet in workshops and seminars, and perhaps almost significantly over coffee and lunch in the Institute's kitchen area; new projects, new collaborations emerge from such encounters. I can offer three examples from the early days of my own experience which have enabled me to keep my own research alive in unexpected ways. All represent new interdisciplinary collaborations.

I met Weisi Guo at the August 2016 'urban priorities' workshop. He presented his work on the analysis of global networks connecting cities which provided the basis for calculating the probabilities of conflicts. This turned out to be of interest to the Ministry of Defence and, through the Institute's partnership in this area, a project to develop this work was funded by DSTL.[11] It seemed to me that Weisi's work could be enhanced by adding flows (spatial interaction) and structural dynamics, and we have worked together on this since our first meeting. New collaborators have been brought in and we have published a number of papers. From each of our viewpoints, adding research from a different, previously unknown field has proved highly fruitful.[12]

The second example took me into the field of health. Mihaela van der Schaar arrived at Turing in October from UCLA, to a Chair in Oxford and a position as a Turing Fellow. One of her fields of research is the application of machine learning to rapid and precise medical diagnosis and prognosis. This is complex territory involving the accounting of co-morbidities as contributing to the diagnosis and prognosis of any particular disease and having an impact on treatment plans. I recognised this as important for the Institute and was happy to support her research. We had a lucky break early on. I gave a breakfast briefing to a group of Chairs and CEOs of major companies. At the end of the meeting I was approached by Caroline Cartellieri. She thanked me for the presentation, but said she wanted to talk to me about something else – she was a Trustee of the Cystic Fibrosis Trust. This led to Mihaela and her team – mainly of PhD students – carrying out a project for the Trust which became an important demonstration of what could be achieved more widely – as well as being valuable for the Trust's own clinicians. For me, it opened up the idea of incorporating the diagnosis and prognosis methods into a 'learning machine' which could ultimately be the basis of personalised medicine. A further thought then occurred. The health learning machine is generic – it can be applied to any flow of people for which there is a possible intervention to achieve an objective. For example, it can be applied to the flow of offenders into and out of prisons; it could be applied in education. Mihaela's methods have also sown the seed of a new approach to urban modelling. The data for the co-morbidities analysis is the record over time of the occurrence of earlier diseases. If these events are re-interpreted in the urban modelling context as 'life events' from demographics, birth, migration and death, but also including entry to education, new job, new house and so on, then a new set of tools can be brought to bear.

The third example, still from very early on, probably autumn 2016, came from me attending for my own education a seminar by Mark Girolami[13] on (I think) the propagation of uncertainty. This is something that I have never been any good at building into urban models. However, I recognised intuitively that his methods seemed to include a piece of mathematics that might possibly solve a problem that has always defeated me – how to predict the distribution of (say) retail centre sizes in a dynamic model. I discussed this with Mark, who enthusiastically agreed to offer the problem to Louis Ellam, then a new research student. He also brought in an Imperial College colleague, Greg Pavliotis, an expert in statistical mechanics and therefore connected to my style of modelling. Over the next couple of years the problem was solved and led to a four-author paper in *Proceedings A of the Royal Society*, with Louis as the first author.[14]

Collaboration in Turing now takes place on a large scale. It has taken me into several fruitful new areas, my collaborators both making the transition manageable and adding new skills. In so doing I have met the '10,000 hours challenge' by proxy.

Pure vs applied

In my early days as CEO in Turing, I was confronted with an old challenge, pure vs applied, though often in a new language: foundational vs consultancy, for example. In my own experiences from schooldays onwards, I was always aware of the higher esteem associated with the 'pure'; indeed, I myself leaned towards the pure end of the spectrum. Even when I started working in physics, I worked in 'theoretical physics'. Only when I converted to the social sciences did I realise that in my new fields I could have it both ways. I worked on the basic science of cities, through mathematical and computer modelling, but with outputs that were almost immediately applicable in town and regional planning and indeed in a commercial context. So where did that kind of thinking leave me in trying to think through a strategy for the Institute?

To oversimplify, there were two camps – the 'foundational' and the 'domain-based'. Some of the former could characterise the latter as 'mere consultancy' and strong feelings were involved. However, there was also a core that straddled the camps – brilliant theorists, able to apply their knowledge in a variety of domains. It was still possible to have it both ways. How might this be turned into a strategy – especially given that the root of a strategic plan would be the allocation of resources to different kinds of research? In relatively early days – it must have been June 2017, when we had the first meeting of our Science Advisory Board – we organised a conference for the second day, inviting the members of our board to give papers. Mike Lynch gave a brilliant lecture on the history of AI through its winters and summers with the implicit question: will the present summer be a lasting one? At the end of his talk, as noted earlier, he said something that has stuck in my mind ever since, 'The biggest challenge for machine learning is the incorporation of prior knowledge,' I would take this further and expand 'knowledge' to 'domain knowledge'. My intuition was that the most important AI and data science research challenges lay within domains, indeed, that the applied problems generated the most challenging foundational problems.

Producing the Institute's Strategic Plan in the context of a sometimes-heated debate proved a long, drawn-out business – it took over a year as I recall. In the end we had a research strategy based on eight challenges, six of which were located in the domains of health, defence and security, finance and the economy, data-centric engineering, public policy and what became 'AI for science'. We had two cross-cutting themes: algorithms and computer science, and ethics. The choice of challenge areas was strongly influenced by our early funders: the Lloyds Register Foundation, GCHQ and the Ministry of Defence, Intel and HSBC. Even without a sponsor at that stage, we could not leave out health. All of these were underpinned by the data science and machine learning methods toolkit. Essentially, this was a matrix structure with columns as domains and rows as methods – an effective way of relaxing the tensions, of having it both ways. This structure has survived, more or less, though new challenges have been added – cities, for example, and the environment.

When it comes to allocating resources, other forces come into play. Do we need some quick wins? The balance between the short term and the longer term is complicated – is the latter inevitably more speculative? Should industry fund most of the applied? This all has to be worked in the context of a rapidly developing government research strategy (with the advent of UKRI)[15] and the development of partnerships with both industry and the public sector. However, there is a golden rule for a research institute (and for many other organisations such as universities, as I learned in my Leeds experience), which is to think through your own strategy rather than simply 'following the money', which is almost always focused on the short term. Then, once the strategy is determined, operate tactically to find the resources to support it.

In making funding decisions, there is an underlying and impossible question to answer about how much has to be invested in an area to produce truly transformative results. This is very much a national question but a version of it exists at the local level. Here is a conjecture: that transformative outcomes in translational areas demand a much larger number of researchers to be funded than to produce such transformations in foundational areas. This is very much for the 'research' end of the R and D spectrum. I can see that the 'D' – development – can be even more expensive. So what did we end up with? The matrix works and at the same time acknowledges the variety of viewpoints. Meanwhile we are continually making judgements about priorities and the corresponding financial allocations. Pragmatism kicks in here.

Leicester City F.C.: a good defence and breakaway goals

Followers of English football will be aware that the top tier is the Premier League and that the clubs that finish in the top four at the end of the season play in the European Champions League the following year. These top four places are normally filled by four of a top half-dozen or so – let's say Arsenal, Chelsea, Liverpool, Manchester City, Manchester United and

Tottenham Hotspur. There are one or two others on the fringe. This group has not traditionally included Leicester City. At Christmas 2014 Leicester were bottom of the Premier League; relegation looked inevitable. Yet they won seven of their last nine games in that season and survived.

At the beginning of the 2015–16 season, the bookmakers' odds on them winning the Premier League were 5000–1 against. Towards the end of that season, they topped the league by eight points with four matches to play, then duly became Premier League champions. The small number of people who bet £10 or more on them at the start of the season, and there were a few, made a lot of money.

How was this achieved? The team had a very strong defence and so conceded few goals; they could score 'on the break', notably through Jamie Vardy, a centre forward who not long ago was playing for Fleetwood Town in the nether reaches of English football; they had an interesting and experienced manager, Claudio Ranieri; and they worked as a team. It was certainly a phenomenon and the bulk of the football-following population were delighted to see them win the League.

What are the academic equivalents? There are university league tables and it is not difficult to identify a top half-dozen. There are tables for departments and subjects. There is a ranking of journals. I do not think there is an official league table of research groups, but there are certainly some informal ones. As in football, it is very difficult to break into the top group from a long way below. Money follows success – as in the REF (the Research Excellence Framework) – and facilitates the transfer of the top players to the top group. So what is the Leicester City strategy for an aspiring university, an aspiring department or research group or a journal editor? The strong defence must be about having the basics in place – good REF ratings and so on. The goal-scoring, break-out attacks are about ambition and risk-taking. The manager can inspire and aspire. As for teamwork, we are almost certainly not as good as we should be in academia, so food for thought there.

There is more, however. All of the above requires at the core hard work, confidence, good plans while still being creative and a preparedness to be different – not to follow the fashion. I am sure that Leicester City had those qualities. So, when *The Times Higher Education Supplement* announces its ever-expanding annual awards, maybe they should add a Leicester City Award for the university that matches their achievement in our own leagues.

Best practice

Everything we do, or are responsible for, should aim at adopting best practice. This is a basic management principle, but easier said than done. We need knowledge, capability and capacity. Then maybe there are three categories through which we can seek best practice: 1) from 'already in practice' elsewhere; 2) could be in practice somewhere but is not because the research has been done but has not been transferred; 3) problem identified, but research needed.

How do we acquire the knowledge? Through reading, networking, Continuing Professional Education (CPE) courses, visits. Capability is about training, experience, acquiring skills. Capacity is about the availability of capability – access to it – for the services (let us say) that need it. Medicine provides an obvious example and local government another. How do each of 164 local authorities in England acquire best practice? Dissemination strategies are obviously important. We should also note that there may be central government responsibilities. We can expect markets to deliver skills, capabilities and capacities through colleges, universities and, in a broad sense, industry itself (in its most refined way through 'corporate universities'). But in many cases there will be a market failure and government intervention becomes essential. In a field such as medicine, which is heavily regulated, the government takes much of the responsibility for ensuring supply of capability and capacity. There are other fields where, in early-stage development, consultants provide the capacity until it becomes mainstream. GMAP in relation to retailing provides an example from my own experience (see pp. 75–7).

How does all this work for cities, in particular for urban analytics? Good analytics provide a better base for decision making, planning and problem solving in city government. This needs a comprehensive information system which can be effectively interrogated. It can be topped with a high-level 'dashboard', with a hierarchy of rich underpinning levels. Warning lights might flash at the top to highlight problems lower down the hierarchy that require further investigation. It needs a simulation (modelling) capacity for exploring the consequences of alternative plans.

Neither of these needs are typically met. In some specific areas it is potentially, and sometimes actually, OK – in transport planning in government, for example, or in network optimisation for retailers. A small number of consultants can and do provide skills and capability. But in general these needs are not met; they are often not even recognised. This seems to be a good example of a market failure. There is central government funding and action through research councils and particularly Innovate UK. The best practice material exists, so we are somewhere in between categories 1 and 2 of the introductory paragraphs above. This tempts me to offer as a conjecture the obvious solution – what is needed are top-class demonstrators. If the benefits were evident, then dissemination mechanisms would follow.

This argument has been presented here in terms of research and its translation into government – the development of best practice in public services, for example. However, it can also be applied to research management itself. We can use the three categories to think through the implications for managing research in a university, or even within a research group.

Time management

When I was Chair of AHRC, I occasionally attended small meetings of academics who we were consulting about various issues – our version of focus groups. On one occasion, we were expecting comments – even complaints – about various AHRC procedures. What we actually heard were strong complaints about the participants' universities who 'did not allow them enough time to do research'. This was a function, of course, of the range of demands in contemporary academic life with at least four areas of work, teaching, research, administration and outreach, all of which figure in promotion criteria. There is a classic time management problem lurking here and the question is whether we can take some personal responsibility for finding the time to do research amidst this sea of demands. More broadly, can the management at a university or research group level find ways of helping the research community with this challenge?

There is a huge literature on time management; I have engaged with it over the years for my own sake as I have tried to juggle a variety of tasks at any one time. The best book I ever found was titled *A-time* by an author whose name I have forgotten – jog my memory please – and which now seems to be out of print. My own copy is long-lost. It was linked to a paper system which helped to deliver its routines. That it is now out of print is probably linked to the fact that I am talking about a pre-PC age. I used that system. I used Filofax. It all helped. There was much sensible advice in the book. 'Do not procrastinate' was particularly good – I still have a problem, but at least I see myself doing something about it. In the pre-email days, correspondence came in the post and piled up in an in-tray; it did not take long for it to form an impossible pile. 'Do not procrastinate' meant deal with it more or less as it comes in. This is true now, of course, of e-mails, which can also quickly get out of hand. I think 'A-time' in the title of the book referred to two things: first, sort out your best and most effective time – morning, night, whatever – your A-time; then divide tasks into A, B and C categories. The trick is to focus your A-time on the A tasks, which makes perfect sense. My up-to-date recommendation on time management is Oliver Burkeman's *Four Thousand Weeks*.[16]

So what does this mean for contemporary academic life? Teaching and administration are relatively straightforward and efficiency is the key. Although sometimes derided, PowerPoint – or an equivalent – is a key aid for teaching once done – no pain, no gain. It can easily be updated (and can also provide the outline of a book).[17] Achieving clarity of expression for different audiences can be very satisfying and creative in its own right. Good writing, as a part of good exposition, is a good training for research writing. So teaching may be straightforward, but it is very important.

Research and outreach are harder. First, research. The choices are harder: what to research, what problem to work on, how to make a difference, how not simply to engage with the pressure to publish for the sake of your CV. (Note the argument in Alvesson's book, *The Triumph of Emptiness*.[18]) So what do we actually do in making research decisions? Here is a mini checklist, summarising earlier argument.

Define your problems – Something interesting and important; interesting at least to you and important to someone else. Be ambitious. Be aware of what others are doing and work out how you are going to be different, not simply a follower of fashion. All easier said than done, of course. And the 'keeping up' is potentially incredibly time-consuming, given the number of journals now current. Perhaps form a 'journals reading club'. All of this is different if you are part of an existing team, but you can – and should – still think as an individual, if only for the sake of your own future.

And finally, outreach. 'Interesting and important' kicks in, but in a different way. Material from both teaching and research can be used. Consultancy becomes possible, though yet another demand on one's time; see pages 75–7.

Thinking things through on all four fronts should produce first a list of pretty routine tasks – administration, 'keeping up' and so on. The rest can be bundled into a number of projects. The two together start to form a work plan with short-run, middle-run and long-run elements. If you want to be very textbook about it, you can define your critical success factors (CSFs), but that may be going too far. So we have a work plan, almost certainly too long and extensive. How do we find the time?

First, be aware of what consumes time: e-mails, meetings, preparing teaching, teaching, supervisions, administration – all of which demand diary management because we have not yet added 'research' to the list. It is important that research is not simply a residual, so time has to be allocated. Within the research box, avoid too much repetition – giving more or less the same paper many times at numerous conferences, for instance. As for outreach, remember to be selective. On all fronts be prepared to use cracks in time to do something useful. In particular, in relation to research, do not wait for the 'free day' or the 'free week' to do the writing. If you have a well-planned outline for a paper, a draft can be written in a sequence of bits of time.

What do I do myself? Am I a paragon of virtue? Of course not, but I do keep a 'running agenda' – a list of tasks and projects with a heading at the top that says 'immediate' and a following one that says 'priorities'. It ends with a list headed 'on the backburner'. Quite often the whole thing is too long and needs to be pruned. When I was at Leeds I used to circulate my running agenda to colleagues because a lot of it concerned joint work of one kind or another. At one point, this ran to 22 pages and, needless to say, I was seriously mocked about it. The lesson is to do it, manage it and control it.

On writing

Research has to be 'written up'. To some writing comes easily, though I suspect this is on the basis of learning through experience. To many, especially research students at the time of thesis writing, it seems like a mountain to be climbed. There are difficulties of getting started and other difficulties of keeping going. An overheard conversation in the centre where I recently worked was reported to me by a third party: 'Why don't you try Alan's 500 words a day routine?' The advice I had been giving to one student – not a party to this conversation – was obviously being passed around.

So let us try that as a starting point. Five hundred words does not feel mountainous. If you write 500 words a day, 5 days a week, 4 weeks a month, 10 months a year, you will write 100,000 words – a thesis, or a long book, or a shorter book and four papers. It is the routine of writing that achieves this, so the next question is how to achieve this routine. This arithmetic, of course, refers to the finished product and this needs preparation. In particular, it needs a good and detailed outline. If this can be achieved, it also avoids the argument that 'I can only write if I have a whole day or a whole week'. The 500 words can be written in an hour or two first thing in the morning; they can be sketched on a train journey. In other words, you can use up bits of time rather than the large chunks that in practice are never available.

The next questions beyond establishing a routine are what to write and how to write. On the first, content is key – you must have something interesting to say. On the second, the most important feature is clarity of expression, which actually reflects clarity of thought. How you do it is for your own voice to determine and that, combined with clarity, will produce your own style.

I can offer one tip on how to achieve clarity of expression – become a journal editor. I was very lucky in that early in my career I became first Assistant Editor of Transportation Research (later TRB) and then Editor of Environment and Planning (later EPA). As an editor you often find yourself in a position of thinking 'There is a really good idea here, but the writing is awful – it does not come through'. You can send it back to the author with suggestions for rewriting but in extreme cases, if the paper is important, you do the rewriting yourself. This process made me realise that my own writing was far from the first rank and I began to edit it as though I was a journal editor. I improved. So the moral can perhaps be stated more broadly – read your own writing as through an editor's eyes with the editor asking, 'What is this person trying to say?'

The content, in my experience, accumulates over time and there are aids to this. First, always carry a notebook. Second, always have a scratchpad next to you as you write to jot down additional ideas that must be squeezed in. The 'how to' is then a matter of having a good structure. What are the important headings? There may be a need for a cultural shift here. Writing at school is about writing essays; it is often the case that a basic principle is laid down which states 'no headings'. I guess this is meant to support good writing so that the structure of the essay and the meaning can be conveyed without headings. I think this is nonsense – though if, say, a magazine demands this, you can delete the headings before submission. This is a battle I am always prepared to fight. In the days when I had tutorial groups, I always encouraged the use of headings. One group refused point blank to do this on the basis of their school principle of 'no headings in essays'. I did some homework and the following week brought in a book of George Orwell's essays, many of which had headings. I argued that if George Orwell could do it, so could everybody. I more or less won.

The headings are the basis of the outline of what is to be written. I would now go further and argue that clarity, especially in academic writing, demands subheadings and sub-subheadings – a hierarchy, in fact. This is now reinforced by the common use of PowerPoint for presentations. This is a form of structured writing and PowerPoint bullets, with sequences of indents, are hierarchical – so we are now all more likely to be brought up with this way of thinking. Indeed, I once had a sequence of around 200 PowerPoint slides for a lecture course. I produced a short book by using this as my outline. I converted the slides to Word, then I converted the now bullet-less text to prose.

I am a big fan of numbered and hierarchical outlines: 1, 1.1, 1.1.1, 1.1.2, 1.2 … 2, 2.1, 2.1.1, 2.1.2 and so on. This is an incredibly powerful tool. At the top level are, say, six main headings, then maybe each has six subheadings and so on. The structure will change as the writing evolves – a main heading disappears and another one appears. This is such a powerful concept that I became curious about who invented it and resorted to Google. There is no clear answer. Indeed, it says something about the contemporary age that most of the references offer advice on how to use this system in Microsoft Word. However, I suspect the origins are probably in Dewey's library classification system, still in use and in effect a classification of knowledge. Google 'Dewey's decimal classification' to find its nineteenth-century history.

There are refinements to be offered on the 'what to' and 'how to' questions. What genre are you working in? Are you producing an academic paper, a book (a textbook?) or a paper intended to influence policy, written for politicians or civil servants? In part, this can be formulated as 'be clear about your audience'. One academic audience can be assumed to be familiar with your technical language; another may be one that you are trying to draw into an interdisciplinary project and might need more explanation. A policy audience probably has no interest in the technicalities, but would like to be assured that they are receiving real 'evidence'.

What next? Start writing, experiment; above all, always have something on the go – a chapter, a paper or a blog piece. Jot down new outlines in that notebook. As Mr Selfridge said, 'There's no fun like work'. Think of writing as fun. It can be very rewarding – when it is finished.

Exercises

1   We should all now believe in lifelong learning. Think about what your own plans might be – in the short, middle and longer term?

2   For a research problem of interest, perhaps one bigger than the one you are working on, how would you put a team together?

3   Pure vs applied: where do you stand – at the poles or in the middle?

4   Consider your response to Exercise 2 above. Is the Leicester City strategy helpful to you?

5   For your research agenda, what constitutes the state of the science? Is that the best practice starting point or do you want to break away?

6   Do you need to take any time management skills on board? Scan the literature and see what you think. A daily mantra for me is 'Do not procrastinate'.

7   Select a paper or a chapter which you need to draft. At the beginning of a week, try the 500 words a day routine, then review what you have by the end of the week. You should be able to do this in a couple of hours a day so that it does not dominate the week.

Notes

1   See Mintzberg, *Structure in Fives*. He argued that if in an organisation the front-line staff were professionals – academics or hospital consultants – then if they were not part of the management, there would be problems. In the UK, universities have done well by this maxim; hospitals have been less successful.

2   The Committee of Vice-Chancellors and Principals (CVCP), now Universities UK (UUK).

3   In my first day as Vice-Chancellor in Leeds, I had to sort out a problem with David Birchall, the Assistant Registrar. I walked down the corridor to his office. The problem was sorted and, as I left, he smiled and said, 'That's a first – the Vice-Chancellor has never been in my office before.' This taught me at first hand something about 'management by walking about'.

4   Mintzberg, *Structure in Fives*.

5   Ashby, *An Introduction to Cybernetics*.

6   Christensen, *The Innovator's Dilemma*.

7   I think this is much-quoted. See Kaplan and Norton, 'The office of strategic management'.

8   For example, Stuart Sutherland, then Principal of Kings College, London, was very helpful.

9   Anthony Finkelstein was the Government Chief Scientific Adviser for National Security. He is now Vice-Chancellor of City University.

10 This was the very popular and much-quoted argument in Gladwell, *Outliers*.

11 The Defence Science and Technology Laboratory.

12 Guo, Gleditch and Wilson (2018).

13 Cambridge Professor and Chief Scientist at The Alan Turing Institute.

14 Ellam, Giriolami, Pavliotis and Wilson, 'Stochastic modelling of urban structure'.

15 UK Research and Innovation (UKRI) took over the responsibility for research funding in the UK from April 2018.

16 Burkeman, *Four Thousand Weeks*.

17 This was indeed how Wilson, *The Science of Cities and Regions* was produced.

18 Alvesson, *The Triumph of Emptiness*.

Chapter 10
Organising research

Introduction

The intention in this chapter is to explore the broader question of the research landscape at bigger scales, for example how the available funding should be allocated. First, however, I consider an example of a field that represents a professional discipline but is in character essentially interdisciplinary – operational research (OR). This involves the application of a range of techniques – essentially modelling – to a wide variety of domains, and therefore combines technical knowledge with domain knowledge. In effect, we pose the 'Weaver question' (see Chapter 7, pp. 92–4) for the future of OR (pp. 127–9). We then turn to the wider question of the allocation of research funding, which in effect means identifying future research priorities. The Weaver question arises again, in this case taking the UK as an example (pp. 129–132). One of the weaknesses of social science research is that it is not funded on the scale that its research challenges demand. We can ask what a CERN for the social sciences would look like (pp. 132–136).

An interdisciplinary discipline: OR in the age of AI

In 2017 I was awarded Honorary Membership of the Operational Research Society. I felt duly honoured, not least because I had considered myself, in part, an operational researcher since the 1970s. I had indeed published in the society's journal and was a Fellow at a time when that had a different status. However, there was a price. The following year I was invited to give the annual Blackett Lecture, delivered to a large audience at the Royal Society in November 2018. The choice of topic was mine. Developments in data science and AI are impacting most disciplines, not least operational research (OR), and I thought that would be an interesting topic to explore. This gave me a snappy title for the lecture – *OR in the Age of AI*. In the context of this book, the lecture and this section raise the question of how to set research priorities in a specific, essentially interdisciplinary field.

OR shares the same enabling disciplines as data science and AI. It is concerned (in outline) with system modelling, optimisation, decision support, planning and delivery. The system's focus forces interdisciplinarity; indeed, this list shows that insofar as it is a discipline, OR shares its field with many others. If we take decision support and planning and delivery as distinguishing OR, at least in part, then we can see it is applied and supports a wide range of customers and clients. These have been through three industrial revolutions and AI promises a fourth. We can think of these customers, public or private, as being organisations driven by business processes. What AI can do is read and write, hear and see, and translate, wonders that will transform many of these business processes. There will be more complicated shifts such as robotics, including soft robotics, a better understanding of markets, increased use of rules-based algorithms to automate processes. Some of these shifts will be large-scale and complicated. In many ways this is all classic OR with new technologies. It is ground-breaking, it is cost saving and it does deplete jobs. In some ways, however, it is not dramatic.

The bigger opportunities come from the scale of available data, computing power and two further things: the ability of systems to learn; and the application of OR to big systems, mainly in the public sector, which are not driven in the way that profit-maximising industries are. For OR this means that the traditional roles of its practitioners will continue, albeit employing new technologies. There is a danger that because these territories overlap across many fields – whether in universities or the big consultancies – many competitors will emerge that could shrink the role of OR. The question then is whether OR can take on leadership roles in the areas of the bigger challenges.

Almost every department of government has these challenges. Indeed many of them, say those associated with the criminal justice system, embrace a range of government departments, each operating in their own silos rather than combining to collect the advantages that could be achieved if they linked their data. All of them have system modelling and/or learning machine challenges. Can OR break into these?

The way to do so is through ambitious proof-of-concept research projects – the 'R' part of R and D – which in turn become the basis for large-scale development projects, the 'D'. There is almost certainly a systemic problem here. Large-scale ambitious projects, usually concerned with building data systems, are arguably a prerequisite and many of these fail. But most of the funded research projects are relatively small and the big 'linking' projects are not tackled. So the challenge for OR, for me, is to open up to the large-scale challenges, particularly in government, and to 'think big'.

The OR community cannot do this alone, of course. However, there is a very substantial OR service in government – one of the recognised analytics professions – and there exists the possibility of asserting more influence from within. But the government itself has a responsibility to ensure that its investment in research is geared to meet these challenges. This has to be a UKRI responsibility – ensuring that research council money is not spread too thinly and that research councils work effectively together, as most of the big challenges are interdisciplinary and cross-council. Government departments themselves should both articulate their own research challenges and be prepared to fund them.

Research and innovation in an ecosystem

UK Research and Innovation (UKRI) is a key node in a complex UK – indeed international – research ecosystem. It can offer strategic direction and for many it will be a key funder.[1]

How are the strategic priorities of a research ecosystem categorised? I am a researcher who has worked in national institutes (the first was the Rutherford Laboratory) and as a university professor, building research teams on research council grants. I was a founder and director of a spin-out company, a university vice-chancellor, Chair of a research council and, most recently, CEO of The Alan Turing Institute. All these activities share common questions and challenges. There is a need to acquire a knowledge of the current landscape and to make decisions on where to invest resources and on how to build capacity and skills. There is also the question of how to connect a top-down strategy with bottom-up creativity. All of these are challenges, on a much bigger scale, for UKRI. Where are the potential game-changers in research? Some will be rooted in pure science; others will be related to wider societal challenges, such as curing cancer. Another key consideration is where knowledge can be applied and this can be used as a working definition of 'innovation'.

How do we set about answering these questions? A systems perspective is valuable, as ever. What is the system of interest and how is it embedded in other systems? The systems view forces an interdisciplinary perspective. At what scale is the research to be focused? Any system of interest will in fact be embedded in a hierarchy of suprasystems and subsystems. Most innovation comes from the lower reaches of the hierarchy;[2] what is more, these discoveries can often be transferred to other domains. Take computers, for example – invented as calculating machines, they are now ubiquitous in a wide range of systems. Contributions to strategy can come top-down from institutions (reading the landscape and horizons scanning) or bottom-up from individual researchers. Impact also plays an important part. Does anyone want to do research that has no impact? I doubt it, but 'impact' should include transformative change in and across disciplines just as much as in industry and the public sector. Perhaps we have been too narrow in our definition of impact.

These challenges, questions and approaches have to be addressed at each node in the ecosystem, after which the nodes must be effectively connected. For example, money has to flow in the direction of the potential game-changers and high-impact innovations. Each node, from the individual piece of research up to UKRI, has to have a strategy, grounded in experience but employing horizon scanning and imagination.

The ecosystem has not been functioning effectively for some time – notably in the transfer of research findings into industry and the public sector. Herein lies a particular challenge for UKRI. Its strategy has to be open to the 'bottom-up' and able to incentivise research councils, Innovate UK, the universities, the research institutes and, last but not least, industry. The ecosystem needs to do all these things if it is to have a chance of delivering game-changers and ground-breaking innovations.

To build an effective strategy, UKRI will have to:

- Identify and build on strengths and opportunities – both established people with track records and the early-career researchers with skills, imagination and ambition. There is a top-down vs bottom-up aspect here.
- Find ways of avoiding the conservatism of peer review which is enforced by the Research Excellence Framework (REF). I believe that universities do not always provide the right incentives, insisting as they do on both the volume of publication and focusing promotion

on 'top journals'. This has skewed the motivation of researchers, particularly by neglecting applied research whose outputs do not qualify for the selected journals.

Industry clearly has a role to play. Where are the modern equivalents of Bell Labs? How much R and D is now being done in start-ups with the big players relying on purchasing success? Although there are many excellent examples of industry-university joint working, there could be many more.

Another strategic question that demands sensitive judgement relates to the size of research groups. What should be located at the 'big science' end of the spectrum? There are established successes, from CERN to Sanger; there are new institutes such as Turing and Diamond, with others in development. But is the average size of a research group in a university too small? Are there potential 'big science' areas that are not (yet) funded as such? Cities, for example, fall into this category. Indeed, how do we value different fields of research for public funding? Health, education, justice – all are obviously important. Basic research is needed to support future industrial development. Should there be more applied research as well, in both industry and the public sector?

We also need to recap on the Weaver question – recall the discussion in Chapter 7 (pp. 81–95). In the 1950s Warren Weaver was the Science Vice-President of The Rockefeller Foundation. He argued that systems of interest fell into three categories: those that were a) simple, b) of disorganised complexity and c) of organised complexity. Roughly speaking, the first two represented (among other things) the physical sciences of the time, while the third comprised biology. Weaver switched his funding from physics to biology, which proved a prescient decision. Is there an equivalent diagnosis to be made now? UKRI's strategy needs to be connected to the social questions of our time: the impacts of climate change and sustainability; the future of work and incomes; our growing social inequalities. Does this agenda demand a Weaver-like shift?

While I have focused on questions specifically relating to UKRI strategy, in reality every element of the research ecosystem needs strategic thinking: from universities and institutes through industry and government departments to individual researchers. All of it needs to be strongly connected to translational and development ecosystems. These challenges are articulated in a report published in 2019 by the British Academy and the Royal Society; it was focused on education research but has wider applications.[3] The underpinning ecosystem concept is vital and three main categories of interest are used: the researchers themselves, the practitioners (more broadly, the users of research) and the policy makers (which could include both those working in a specific area such as education and those concerned with research policy). The ecosystem players create a variety of demands for research. The researchers and some associated policy makers might focus on the 'blue skies' dimension. The users will have more or less well-articulated demands. The policy makers in a domain will want to be science-led (that is, research-led). The research policy makers will face the challenge of allocating funds in a situation where resources are scarce and cannot possibly meet all the demand.[4]

In conclusion, we can return to the core argument implicit throughout this book, which is that most research challenges are by their nature interdisciplinary. The research ecosystem still has too many of its most powerful nodes rooted in disciplines. A major restructuring challenge is required to transfer the balance from the 'within-discipline' to interdisciplinary foundations.

Scale: a CERN for the social sciences?

There is a broad issue here of support for the social sciences relative to the 'hard' sciences. This can be illustrated in relation to cities. Urban and regional modelling (or, more broadly, analytics) is the foundation of a science of cities and this underpins approaches to the associated planning challenges. The field as we know it has almost 60 years of history.[5] A wide variety of approaches exist, rooted in several different disciplines. As a result the field is very fragmented; there is, as yet, no 'normal' science in Kuhn's sense.[6] This is partly because the science of cities is a young field and partly because it has never been a 'big science' – in contrast to elementary particle physics with CERN, for example, the epitome of a 'big' international collaboration. Relative to its potential contribution to problem solving, research on cities – and the social sciences more broadly – has been under-resourced. However, some of the fragmentation arises because of modellers' tendencies to leap into disciplinary camps (or silos); each camp then argues that their method is 'best', or at least fails to engage with alternatives. The first principle to be adopted, therefore, is 'no more silos'. A corollary of this principle is to extend horizons – cross and link scales, cross and link disciplines, work on previously neglected problems.[7] This all demands research on a bigger scale.

A second principle, stated rather subjectively, but to be illustrated below, is to make sure that the elements of this research are 'interesting' and 'important'. The science should be interesting, the applications important. A lot of progress has been made in developing the science, albeit in a fragmented form. Much less progress has been made in using this science to support various forms of planning. There are significant exceptions as we have seen – transport in the public sector and retail in the private sector, for instance. Yet there seems to be a profound disconnect between the modelling and planning communities. This is true at an elementary, basic level and also, perhaps less surprisingly, in relation to the most difficult or 'wicked problems'. A corollary of this principle, therefore, is that at least part of the modelling community should ensure that they are fully contributing to planning problems of all kinds; the planning community should welcome this help and possibly pay for (some of) it. In relation to both principles, there is a need to be ambitious.

What should a utopian CERN-like institute (or network of institutes) be doing? I will illustrate the argument with examples from my own experience and prospective future research, though this is a small contribution. Potentially, this kind of thinking opens up a very extensive agenda.

We can now take for granted computing power, high levels of visualisation skills and data in abundance through the 'big data' movement. We have an extensive existing 'toolkit'.[8] We also have excellent, account-based models of demographics and economics; we can model the functioning of most subsystems, though these skills are not universally applied; and we have the beginnings of an understanding of dynamics, with its implications of path dependence and phase changes. There are in addition 'new kids on the block' with, for example, agent-based modelling. So what are the key ongoing challenges? Examples are:

- Interdisciplinary integration at the research front line.
- The need to deal with the high dimensionality of the algebraic arrays that constitute our models – which I would argue means the full integration of microsimulation with other methods.
- The need to gain a fuller understanding of structure-generating mathematics. At the moment we are good on approaches which build on Lotka-Volterra models and the beginnings of Turing-style reaction-diffusion models, but not on much else.
- Articulation of 'best practice' so that the kit is fit for purpose in policy making and planning.

We then need to move on to the challenges in planning, particularly the 'wicked problems'. Examples (illustrated in the context of England, but noting that there will be similar lists for other countries) are:

- Economic development at the urban and regional scale – routine application of input-output models at these scales, understanding the future of (un)employment, impacts of changing technologies.
- Education – failing schools, parental choice.
- Health – evaluating structural reorganisation in attempts to solve the problems of the NHS.
- The welfare benefits system.
- Housing – the shift to rental, homelessness.
- Police reform.
- The criminal justice system, prison reform, prisoner rehabilitation.
- Joining up – multiple deprivation.
- Inner-city regeneration.
- Major UK infrastructure planning – moving beyond shopping lists.

In all of these areas, although we have the modelling toolkit which could be used at least for short-run analysis, there is very little modelling work. The modellers' agenda has been too narrow. Indeed, it could be argued that these 'wicked problems' could be treated as operational research. (Is it the existence of subdisciplinary silos that means operational researchers and urban modellers now barely intersect?) The policy and planning communities are also unreceptive.

We need comprehensive models with all the subsystem models stitched together. We might then begin to see the extent to which problems such as poor health, housing and unemployment are essentially income problems, that these are access-to-work issues and that these in turn are skills issues. Is the education system the key? The answer is probably 'yes', but we have not analysed it in these terms. We could use models representing the middle and longer runs to estimate, on the basis of alternative policies and plans, how long it would take to have any serious impact on this kind of agenda.

It needs to be emphasised, of course, that few of the 'wicked problems' can be solved through the kind of capital investment and structural reform for which models provide a good base for evaluation. In most cases there are also deep problems of culture and working practices. These require different kinds of analysis and policies to be developed in parallel.

The stitching together, more broadly, needs to be accomplished within government – both local and national. The data is available; information, understanding and knowledge can be generated through the application of models. Plans can be designed and test results recorded. The joining up would be through what might be called an IGIS – an intelligent geographical information system. Such an information system could house the GIS-type data as now. However, it would be fundamentally enhanced by the addition of plans and policies and tests of these through being able to call up appropriate model runs and visualising and evaluating the outputs.

It is clear that not all of this can be done within my utopian institute, however large, although it could lead by developing best practice and articulating the research front line. Government agencies would have to play their part with substantial in-house integrated units. Universities would also have a major role with seriously large interdisciplinary centres. Each of these needs to be substantially bigger than typical core departments, so a major restructuring is called for. Inevitably, battles would be waged as disciplines went into defensive mode. The vision must be for an extended utopia.

Utopias, by definition, are not achievable. However, it is possible to use them to chart new directions. Local government should have cross-council units that would construct an intelligent information system for a city or region; such units could incorporate basic data through a GIS, policies and plans, and a model-based analysis and evaluation system. A best-practice system is needed to chart the way. In the UK an example was provided through the then Technology Strategy Board[9] – a kind of research council for innovation. This funded a £24M project for Glasgow as a 'future cities demonstrator', specifically to show the benefits of integrating data and services.

More broadly, the UK Ministry of Communities and Local Government[10] should lead a cross-department joining-up initiative. Universities need to take interdisciplinary challenges more seriously and some, at least, should use cities and regions as a case study of what can be achieved. They would need to invest themselves, but would also require partners from among research councils, consultants and government agencies. A new kind of co-operative structure is needed here. There are significant implications for education too; new courses will be needed for both young and mature students to generate an expanding skilled workforce in modelling and planning. The modelling community should break out of silos, extend its horizons and work in the major public services (and the private sector), possibly re-invigorating operational research en route. This is non-trivial and, with some notable exceptions, it is not what we have now. The policy and planning communities need to embrace intelligent analysis.

This all turns on a commitment to ambition in both the science and in policy and planning. The science needs to become 'big science' – exciting in its own terms, but able to contribute to solving some of society's biggest problems. This is not essentially a funding problem: universities and government agencies could reorganise and commit finance from existing resources. It is rather a cultural problem with many dimensions. For example, academic researchers, in inefficient silos, have become accustomed to working on small problems, often on 'toy' systems, because they are manageable; policy makers and planners typically have short-term perspectives and lack the backgrounds to see that the science can help. Many are also unambitious in that part of their own territory they should be good at – inventing and developing radical plans for problems that need radical solutions. These cultural issues are deep-seated, but dreaming the utopian dream could perhaps be the beginning of something new. Brian Arthur's argument shows that if the breaking down of the boundaries of silos and the extension of horizons in both modelling and planning could be achieved, combinatorial evolution would then ensure more rapid progress than we can at present envisage.

Exercises

1   Think of examples, such as OR, of subdisciplines that need to re-adjust to meet future challenges.
2   Outline the ecosystem of research and associated funding of the field of your own research project.

Notes

1   UKRI – UK Research and Innovation – is the UK body that allocates government research funding.
2   Arthur, *The Nature of Technology*, Chapter 2.
3   Royal Society, *Harnessing Educational Research*.
4   There is an argument here for multiplicity of funding sources – as does exist, of course, since there cannot be a single omniscient body.
5   This was brought home to me as I edited five volumes of the history of urban modelling. See Wilson, ed., *Urban modelling*.
6   T. Kuhn (1962) *The structure of Scientific Revolutions*, University of Chicago Press, Chicago
7   Although the argument here relates to contemporary problems, there are enormous opportunities for modellers in areas such as history and archaeology. For a toe in the water see Bevan and Wilson, 'Models of settlement hierarchy based on partial evidence'.
8   See Wilson, *The Science of Cities and Regions*.
9   Now Innovate UK, part of UKRI (see Note 1 above).
10   Retitled in 2021 to reflect the 'levelling up' agenda with a cross-government role.

Bibliography

Ackoff, R. L. *Ackoff's Best: His classic writings on management*. New York: Wiley, 1999.

Aleksander, I. and H. Morton. *An Introduction to Neural Computing*. London: Chapman and Hall, 1990.

Alexander, C. *Notes on the Synthesis of Form*, Cambridge, MA: Harvard University Press, 1964.

Alvesson, M. *The Triumph of Emptiness*. Oxford: Oxford University Press, 2013.

Andersson, C. *Urban Evolution*. Goteborg: Department of Physical Resource Theory, Chalmers University of Technology, Goteborg, 2005.

Anderson, P. W., K. J. Arrow and D. Pines, eds. *The Economy as an Evolving Complex System*. Menlo Park, CA: Addison Wesley, 1988.

Angier, N. *The Canon: The beautiful basics of science*. London: Faber and Faber, 2007.

Arthur, W. B. 'Urban systems and historical path dependence'. In *Cities and their Vital Systems: Infrastructure, past, present and future*, edited by J. H. Ausubel and R. Herman, R., 85–97. Washington, DC: National Academy Press, 1988.

Arthur, W. B. *Increasing Returns and Path Dependence in the Economy*. Ann Arbor, Michigan: University of Michigan Press, 1994a.

Arthur, W. B. 'Inductive reasoning and bounded rationality', *American Economic Association Papers and Proceedings* 84 (1994b): 406–11.

Arthur, W. B. *The nature of technology*. New York: The Free Press, 2009.

Ashby, W. R. *An Introduction to Cybernetics*. London: Chapman and Hall, 1956.

Bailey, F. G. *Morality and Expediency: The folklore of academic politics*. Oxford: Blackwell, 1977.

Barber M. *How to Run a Government so that Citizens Benefit and Taxpayers Don't Go Crazy*. Harmondsworth, Middx: Penguin, 2016.

Batty, M. and Milton, R. 'A new framework for very large-scale urban modelling', *Urban Studies* 58 (2021). *DOI:10.1177.0042098020982252*.

Baudains, P., S. Zamazalová, M. Altaweel and A. G. Wilson. 'Modeling strategic decisions in the formation of the early Neo-Assyrian Empire', *Cliodynamics: The Journal of Quantitative History and Cultural Evolution* 6(1) (2015): 1–23.

Baudains, P. and A. Wilson. 'Conflict Modelling: Spatial interaction as threat'. In *Global Dynamics*, edited by A. Wilson, 145–58. Chichester: Wiley, 2016.

Becher, T. *Academic Tribes and Territories*. Milton Keynes: Open University Press, 1989.

Beck, C. and F. Schlogl. *Thermodynamcs of Chaotic Systems*. Cambridge: Cambridge University Press, 1993.

Becker, H. *Tricks of the Trade: How to think about your research while you're actually doing it*. Chicago: University of Chicago Press, 1998.

Beer, S. *Brain of the Firm*. Chichester: Wiley, 1972 (second edition 1981).

Beer, S. *Designing Freedom*. Chichester: Wiley, 1994.

Bertalanffy, L. von. *General System Theory*. New York: Braziller, 1968.

Bevan, A. and A. G. Wilson. 'Models of settlement hierarchy based on partial evidence', *Journal of Archaeological Science* 40(5) (2013): 2415–27.

Birkin, M., G. P. Clarke, M. Clarke and A. G. Wilson. *Intelligent GIS: Location decisions and strategic planning*. Cambridge: Geoinformation International, 1996.

Blum, B. I. *Beyond Programming: To a new era of design*. Oxford; New York: Oxford University Press, 1996.

Bochel, H. and S. Duncan, eds. *Making Policy in Theory and Practice*. Bristol: The Policy Press, 2007.

Bok, D. *Higher Learning*. Cambridge, MA: Harvard University Press, 1986.

Boltzmann, L. *Lectures on Gas Theory* [1896]. Berkeley and Los Angeles: University of California Press, translated by S. G. Brush, 1964.

Brouwer, L. E. J. 'Uber eineindeutige stige Transformationen von Flachen in Sich', *Mathematische Annalen* 67 (1910): 176–80.

Boyce, D. E. and H. Williams. *Forecasting Urban Travel: Past, present and future*. Cheltenham and Northampton, MA: Edward Elgar Publishing, 2015.

Burkeman, O. *Four Thousand Weeks*. London: Bodley Head, 2021.

Buzan, T. with Buzan, B. *The Mind Map Book*. London: BBC Publications, 1993.

Chapman, G. T. *Human and Environmental Systems*. London: Academic Press, 1977.

Checkland, P. *Systems Thinking, Systems Practice*. Chichester: Wiley, 1981.

Checkland, P. and J. Scholes. *Soft systems methodology in action*. Chichester: Wiley, 1991.

Checkland, P. and S. Holwell. *Information, Systems and Information Systems*. Chichester: Wiley, 1998.

Christaller, W. *Die centralen Orte in Suddeutschland*. Jena: Gustav Fisher, 1933; English translation by C. W. Baskin, *Central Places in Southern Germany*. Englewood Cliffs, N. J.: Prentice Hall, 1966.

Christensen, C. M. *The Innovator's Dilemma*. New York: Harper Business, 1997.

Cisco. *Equipping Every Learner for the 21st Century*. San Jose, CA: Cisco Systems Inc, 2007.

Clarke, B. R. *Creating Entrepreneurial Universities: Organizational pathways of transformation*. Oxford: Pergamon, 1998.

Clarke, C., ed. *The Too Difficult Box*. London: Biteback Publishing, 2014.

Clarke, G. P., ed. *Microsimulation for Urban and Regional Policy Analysis*. London: Pion, 1996.

Clarke, G. P. and A. G. Wilson. 'Performance indicators and model-based planning I: The indicator movement and the possibilities for urban planning', *Sistemi Urbani* 2 (1987a): 79–123.

Clarke, G. P. and A. G. Wilson. 'Performance indicators and model-based planning II: Model-based approaches', *Sistemi Urbani* 9 (1987b): 138–65.

Clarke, M. *How Geography Changed the World, and My Small Part in It*. Bristol: Sweet Design Ltd, 2020.

Cromer, A. *Connected Knowledge: Science, Philosophy and Education*. Oxford: Oxford University Press, 1997.

Cronon, W. *Nature's Metropolis*. New York: Norton, 1991.

Dantzig, G. B. *Linear Programming and Extensions*. Princeton, N.J.: Princeton University Press, 1963.

Davies, T., H. Fry, A. G. Wilson and S. R. Bishop. 'A mathematical model of the London riots and their policing', *Nature Scientific Reports* 3 (1303) (2013). doi:10.1038/srep01303.

Davies, T., H. Fry, A. Wilson, A. Palmisano, M. Altaweel and K. Radner. 'Application of an entropy maximizing and dynamics model for understanding settlement structure: The Khabur Triangle in the Middle Bronze and Iron Ages', *Journal of Archaeological Science* 43 (2014). doi:10.1016/j.jas.2013.12.014.

Dearden, J., Y. Gong, M. Jones and A. Wilson. 'Using state space of a BLV retail model to analyse the dynamics and categorise phase transitions of urban development', *Urban Science* 3 (2019): 31–47. doi.10.3390/urbansci.3010031.

Dearden, J. and A. G. Wilson. 'A framework for exploring urban retail discontinuities', *Geographical Analysis* 43 (2) (2011), 172–87.

Dearden, J. and A. G. Wilson. 'The relationship of dynamic entropy maximising and agent-based approaches'. In *Urban Modelling in Agent-Based Models of Geographical Systems*, edited by A. J. Heppenstall, A. T. Crooks, A. T. See and L. M. Batty, Chapter 35, 705–20. Berlin: Springer, 2011.

Dearden, J. and A. G. Wilson. *Explorations in Urban and Regional Dynamics*. Abingdon: Routledge, 2015.

Dennett, A. and A. G. Wilson. 'A multi-level spatial interaction modelling framework for estimating interregional migration in Europe', *Environment and Planning A* 45 (2013): 1491–507.

Dijkstra, E. W. 'A note on two problems in connection with graphs', *Numerische Mathematik* 1 (1959): 269–71.

Drucker, P. F. *The New Realities*. London: Heinemann, 1989.

Ellam, L., M. Girolami, G. Pavliotis and A. Wilson. 'Stochastic modelling of urban structure', *Proceedings of the Royal Society A* 474: 20170700, http://dx.doi:10.1098/rspa.2017.0700.

Epstein., J. M. and R. Axtell. *Growing Artificial Societies: Social science from the bottom up.* Cambridge, MA: MIT Press, 1996.

Epstein, J. M. *Nonlinear Dynamics, Mathematical Biology and Social Science.* Reading, MA: Addison-Wesley, 1997.

Evans, S. P. 'A relationship between the gravity model for trip distribution and the transportation model of linear programming', *Transportation Research* 7 (1973): 39–61.

Feynman, R. P., R. B. Leighton and M. Sands, M. *The Feynman Lectures on Physics.* Reading, MA: Addison-Wesley, 1963.

Foster, C. D. *British Government in Crisis: The third English revolution.* Oxford; Portland, Oregon: Hart Publishing, 2005.

Foster, C. D. and Beesley, M. E. 'Estimating the social benefit of constructing an underground railway in London', *Journal of the Royal Statistical Society A* 126 (1963): 46–92.

Gibbons, M., C. Limoges, H. Nowotny, S. Schwartzman, P. Scott and M. Trow. *The New Production of Knowledge: The dynamics of science and research in contemporary societies.* London: Sage, 1994.

Gladwell, M. *Outliers: the strategy of success.* New York: Little Brown, 2008.

Glass, N. M. *Management Masterclass: A practical guide to the new realities of business.* London: Nicholas Brealey, 1996.

Government Office for Science. 'Future of Cities project reports', 2013:
a) Overview: https://www.gov.uk/government/uploads/system/uploads/attachment_data/file/520963/GS-16-6-future-of-cities-an-overview-of-the-evidence.pdf. Accessed 3 January 2022; b) Science of Cities: https://www.gov.uk/government/uploads/system/uploads/attachment_data/file/516407/gs-16-6-future-cities-science-of-cities.pdf. Accessed 3 January 2022; c) Foresight for Cities: https://www.gov.uk/government/uploads/system/uploads/attachment_data/file/516443/gs-16-5-future-cities-foresight-for-cities.pdf. Accessed 3 January 2022; d) Graduate Mobility: https://www.gov.uk/government/uploads/system/uploads/attachment_data/file/510421/gs-16-4-future-of-cities-graduate-mobility.pdf. Accessed 3 January 2022.

Guo, W., K. Gleditsch and A. Wilson (2018) 'Retool AI to forecast and limit wars', *Nature* 562 (2018): 331–3.

Habermas, J. *Theory and Practice.* London: Heinemann, 1974.

Haggett, P. *Locational Analysis in Human Geography.* London: Edward Arnold, 1965.

Hall, P. *Cities of Tomorrow.* Oxford: Blackwell, 1988.

Hall, W. and J. Pesenti, J. *Growing the Artificial Intelligence Industry in the UK.* London: DCMS (Department for Culture, Media and Sport) and BEIS (Department for Business, Energy and Industrial Strategy), 2017.

Hamilton, I. *Against Oblivion.* London: Viking Penguin, 2002.

Hansen, W. G. 'How accessibility shapes land use', *Journal of the American Institute of Planners* 25 (1959): 73–6.

Harris, B. (1965) 'Urban development models: New tools for planners', *Journal of the American Institute of Planners* 31 (1965): 90–5.

Harris, B. and Wilson, A. G. (1978) 'Equilibrium values and dynamics of attractiveness terms in production-constrained spatial-interaction models', *Environment and Planning A*, 10 (1978): 371–88.

Harvey, D. *Social Justice and the City.* Baltimore. ML: Johns Hopkins University Press, 1973.

Hawken, P. *The Ecology of Commerce.* New York: Harper-Collins, 1993.

Herbert, D. J. and Stevens, B. H. 'A model for the distribution of residential activity in an urban area', *Journal of Regional Science* 2 (1960): 21–36.

Hidalgo, C. *Why Information Grows.* Harmondsworth, Middx: Penguin, 2015.

Holland, J. H. *Adaptation in Natural and Artificial Systems: An introductory analysis with applications in biology, control and artificial intelligence.* Cambridge, MA: MIT Press, 1992.

Holland, J. H. *Hidden Order: How adaptation builds complexity.* Reading, MA: Addison-Wesley, 1995.

Holland, J. H. *Emergence.* Reading, MA: Addison-Wesley, 1998.

Horgan, J. *The End of Science.* London: Little Brown and Co. (UK), 1997.

Hotelling, H. 'Stability in competition', *Economic Journal* 39 (1929): 41–57.

Hudson, L. *The Cult of the Fact.* London: Jonathan Cape, 1972.

Isard, W. *Location and the Space-Economy.* Cambridge, MA: MIT Press, 1956.

Isard, W. *Methods of Regional Analysis.* Cambridge, MA: MIT Press, 1960.

Jacobs, J. *The Economy of Cities*. London: Jonathan Cape, 1970.

Jaynes, E. T. 'Information theory and statistical mechanics', *Physical Review* 106 (1957): 620–30.

Jaynes, E. T. *Probability Theory*. Cambridge: Cambridge University Press, 2003.

Johnson, G. *University Politics: F. M. Cornford's Cambridge and his advice to the young academic politician*. Cambridge: Cambridge University Press, 1994.

Johnson, S. *Where Good Ideas Come From: The seven patterns of innovation*. London: Penguin, 2010.

Jordan, M. I. 'Artificial intelligence: The revolution hasn't happened yet', *Harvard Data Science Review* 7(1) (2019). https://doi.org/10.1162/99608f92.f06c6e6.

Kaplan, R. S. and D. P. Norton. 'The office of strategic management', *Harvard Business Review*, 2005.

Kim, T. J., D. E. Boyce and G. J. D. Hewings. 'Combined input–output and commodity flow models for interregional develoment planning', *Geographical Analysis* 15 (1983): 330–42.

Klir, J. and M. Valach. *Cybernetic Modelling*. London: Illiffe, 1967.

Kostitzin, V. A. *Mathematical Biology*. London: Harrap, 1939.

Kronman, A. T. *Education's End: Why our colleges and universities have given up on the meaning of life*. New Haven, CT: Yale University Press, 2007.

Kuhn, T. (1962) *The Structure of Scientific Revolutions*, University of Chicago Press, Chicago

Lakoff, G. and M. Johnson. *Metaphors We Live By*. Chicago: University of Chicago Press, 1980.

Lakshmanan, T. R. and W. G. Hansen. 'A retail market potential model', *Journal of the American Institute of Planners* 31 (1965): 134–43.

Lancaster, K. 'A new approach to consumer theory', *The Journal of Political Economy* 74 (1966): 132–57.

Lancaster, K. J. *Consumer Demand: A new approach*. New York: Columbia University Press, 1971.

Lee, D. B. 'Requiem for large-scale models', *Journal of the American Institute of Planners* 39 (1973): 163–78.

Leontief, W. *Input-Output Analysis*. Oxford: Oxford University Press, 1967.

Lotka, A. J. *The Elements of Physical Biology*. Baltimore, ML: Williams and Wilkins, 1925.

Lowry, I. S. *A Model of Metropolis*. Santa Monica: RM-4035-RC, The Rand Corporation, 1964.

McCann, P. *Urban and Regional Economics*. Oxford: Oxford University Press, 2001.

Madge, J., G. Colavizza, J. Hetherington, W. Guo and A. Wilson. 'Assessing simulations of imperial dynamics and conflict in the ancient world', *Cliodynamics* 10(2) (2019): 25–39.

May, R. M. 'Stability in multi-species community models', *Mathematical Biosciences* 12 (1971): 59–79.

May, R. M. *Stability and Complexity in Model Ecosystems*. Princeton, N.J.: Princeton University Press, 1973.

McIrvine E.C. and M. Tribus (1971) 'Energy and information', *Scientific American*, 225 (3), 179–90

Medda, F. R., F. Caravelli, S. Caschili and A. Wilson. *Collaborative Approach to Trade: Enhancing connectivity in sea- and land-locked countries*. Heidelberg: Springer, 2017.

Mintzberg, H. *Structure in Fives: Designing effective organisations*. Englewood Cliffs, N.J.: Prentice Hall, 1989.

Mullan, J. *How Novels Work*. Oxford: Oxford University Press, 2006.

Nadler, D. A., M. S. Gerstein, R. B. Shaw and associates. *Organisational Architecture: Designs for changing organisations*. San Francsico, CA: Jossey-Bass, 1992.

Nash, J. 'Equilibrium points in n-person games', *Proceedings of the National Academy of Sciences* 36 (1950): 48–9.

National Academy of Sciences, National Academy of Engineering and Institute of Medicine. (2007) *Rising Above the Gathering Storm: Energizing and employing America for a brighter economic future*. Washington D.C.: The National Academies Press, 2007.

National Infrastructure Commission. *Data for the Public Good*. 2018.

Neumann, J. von. *Theory of Self-Reproducing Automata*. Urbana, IL: University of Illinois Press, 1966.

Newman, M., A-L. Barabasi and D. J. Watts, D. J. *The Structure and Dynamics of Networks*. Princeton, N.J.: Princeton University Press, 2006.

Nicolis, G. and I. Prigogine. *Self-Organisation in Non-Equilibrium Systems: From dissipative structures to order through fluctuations*. Chichester: Wiley, 1977.

Nowak, M. A. *Evolutionary Dynamics: Exploring the equations of life*. Cambridge, MA: Belknap Press of Harvard University Press, 2006.

Nowak, M. A. and R. M. May. *Virus Dynamics: Mathematical principles of immunology and virology*. Oxford: Oxford University Press, 2000.

Nystuen, J. D. and M. F. Dacey. 'A graph theory interpretation of nodal regions', *Papers, Regional Science Association* 7 (1961): 29–42.

Orcutt, G. H. 'A new type of socio-economic system', *Review of Economic Statistics* 58 (1957): 773–97.

Pagliara, F., M. de Bok, D. Simmonds and A. G. Wilson, eds. *Employment Location in Cities and Regions: Models and applications*. Heidelberg: Springer, 2012.

Papdimitrious, C. H. and Steiglitz, K. *Combinatorial Optimization: Algorithms and complexity*. Englewood Cliffs, N.J.; Prentice Hall (Dover edition), 1998.

Polya, G. *How to Solve It*. Princeton, N.J.: Princeton University Press, 1945. Quandt, R. E. and W. J. Baumol. 'The demand for abstract modes: Theory and measurement', *Journal of Regional Science* 6 (1966): 13–26.

Rees, P. H. and A. G. Wilson. *Spatial Population Analysis*. London: Edward Arnold, 1976.

Rihll, T. E. and A. G. Wilson. 'Spatial interaction and structural models in historical analysis: Some possibilities and an example', *Histoire et Mesure II-1* (1987a): 5–32.

Rihll, T. E. and A. G. Wilson. 'Model-based approaches to the analysis of regional settlement structures: The case of ancient Greece'. In *History and Computing*, edited by P. Denley and D. Hopkin, 10–20. Manchester: Manchester University Press, 1987b.

Rihll, T. E. and A. G. Wilson. 'Settlement structures in Ancient Greece: New approaches to the polis'. In *City and Country in the Ancient World*, edited by J. Rich and A. Wallace-Hadrill, 58–95. London: New York: Routledge.

Richardson, L. F. *Arms and Insecurity*. Pittsburgh: The Boxwood Press, 1960.

Rittel, H. W. J. and M. M. Webber. 'Dilemmas in a general theory of planning', *Policy Sciences* 4 (1973): 155–69.

Robinson, K. *The Element: How finding your passion changes everything*. London: Penguin, 2009.

Royal Society. *Harnessing Educational Research*. London: British Academy and the Royal Society, 2019.

Rosser, J. B. Jr. *From Catastrophe to Chaos: A general theory of economic discontinuities*. Boston: Kluwer Academic Publishers, 1991.

Roumpani, F. PhD thesis, University College London. 2018.

Rudlin, D. and N. Falk. 'Uxcester Garden City', Wolfson Prize, URBED. Manchester, 2014.

Ruelle, D. *Chance and Chaos*. Hamondsworth, Middx: Penguin, 1991.

Ruelle, D. *The Thermodynamic Formalism: The mathematical structure of equilibrium statistical mechanics*. Cambridge: Cambridge University Press, 2002.

Sadler, P. *Designing Organisations*. London: Mercury Books, 1991.

Scarf, H. *The Computation of Economic Equilibria*. New Haven: Yale University Press, 1973a.

Scarf, H. 'Fixed-point theorems and economic analysis', *American Scientist* 71 (1973b): 289–96.

Schlecty, P. C. *Creating Great Schools: Six critical systems at the heart of educational innovation*. San Francisco, CA: Jossey-Bass, 2005.

Sen, A. *Commodities and Capabilities*. Oxford: Oxford University Press, 1999.

Sennett, R. *The Corrosion of Character: The personal consequences of work in the new capitalism*. New York: W. W. Norton, 1998.

Sennett, R. *The Culture of the New Capitalism*. New Haven, CT: Yale University Press, 2006.

Sennett, R. *The Craftsman*. London: Penguin, 2008.

Shannon, C. and Weaver, W. *The Mathematical Theory of Communication*. Urbana, IL: University of Illinois Press, 1949.

Simmonds, D. 'The design of the DELTA land-use modelling package', *Environment and Planning B* 26 (1999): 665–84.

Simon H. A. (1956) 'Rational choice and the structure of the environment', *Psychological Review*, 63, 129–138

Simon H. A. *The Sciences of the Artificial*. Cambridge, MA: MIT Press, 1996 (third edition).

Singleton, A. D., A. G. Wilson, A. G. and O. O'Brien. 'Geodemographics and spatial interaction: An integrated model for higher education', *Journal of Geographic Systems* 14:223–41. doi: 10.1007/s10109.010-0141-5, Online First.

Stern, N. *The Economics of Climate Change*. Cambridge: Cambridge University Press, 2007.

Stewart, I. *Seventeen Equations that Changed the World*. London: Profile Books, 2012.

Stone, R. *Mathematics in the Social Sciences*. London: Chapman and Hall, 1967.

Stone, R. *Mathematical Models of the Economy*. London: Chapman and Hall, 1970.

Swinney, P. and E. Thomas. 'A century of cities: Urban economic change since 1911', *Centre for Cities*, available at: https://www.centreforcities.org/wp-content/uploads/2015/03/15-03-04-A-Century-of-Cities.pdf. Accessed 3 January 2022.

Thom, R. *Structural Stability and Morphogenesis*. Reading, MA: W. A. Benjamin, 1975.

Thomsen, E. *OLAP Solutions: Building multidimensional information systems*. New York: Wiley, 1997.

Thunen, J. H. von. *Der isolierte staat in beziehung auf landwirtschaft und nationalokonomie*. Stuttgart: Gustav Fisher, 1826. English translation (*The Isolated State*) by C. M. Wartenburg. Oxford: Oxford University Press, 1966.

Tolstoy, L. *War and Peace* [1868–9]. Harmondsworth, Middx: Penguin, 2005, 915.

Turing, A. M. 'The chemical basis of morphogensis', *Philosophical Transactions of the Royal Society of London, series B* 237 (1952): 37–72.

Volterra, V. 'Population growth, equilibria and extinction under specified breeding conditions: A development and extension of the theory of the logistic curve', *Human Biology* 10 (1938): 173–8.

Weaver, W. 'Science and complexity', *American Scientist* 36 (1948): 536–44.

Weaver, W. 'A quarter century in the natural sciences', *Annual Report (1958)*. New York, The Rockefeller Foundation, 7–122.

Weiner, N. *Invention*. Cambridge, MA: MIT Press (1994 edition).

Williams, H. C. W. L. 'On the formation of travel demand models and economic evaluation measures of user benefit', *Environment and Planning* A 9 (1977): 285–344.

Wilson, A. G. 'A statistical theory of spatial distribution models', *Transportation Research* 1 (1967): 253–69.

Wilson, A. G. *Entropy in Urban and Regional Modelling*. London: Pion, 1970.

Wilson, A. G. 'Generalising the Lowry model', *London Papers in Regional Science* 2 (1971): 121–34.

Wilson, A. G. *Urban and Regional Models in Geography and Planning*. Chichester; New York: Wiley, 1974.

Wilson, A. G. 'Spatial interaction and settlement structure: Towards an explicit central place theory'. In *Spatial Interaction Theory and Planning Models*, edited by A. Karlquist, L. Lundquist, F. Snickars and J. W. Weibull, 137–56. Amsterdam: North Holland, 1978.

Wilson, A. G. *Catastrophe Theory and Bifurcation: Applications to urban and regional systems*. London: Croom Helm; Berkeley, CA: University of California Press, 1981.

Wilson, A. G. *Complex Spatial Systems: The modelling foundations of urban and regional analysis*. Harlow: Prentice Hall, 2000a.

Wilson, A. G. 'The widening access debate: Student flows to universities and associated performance indicators', *Environment and Planning A* 32 (2000b): 2019–31.

Wilson, A. G. 'Ecological and urban systems models: Some explorations of similarities in the context of complexity theory', *Environment and Planning A* 38 (2006): 633–46.

Wilson, A. G. 'A general representation for urban and regional models', *Computers, Environment and Urban Systems* 31 (2007): 148–61.

Wilson, A. G. 'Boltzmann, Lotka and Volterra and spatial structural evolution: An integrated methodology for some dynamical systems', *Journal of the Royal Society, Interface* 5 (2008): 865–71. *doi:10.1098/rsif.2007.1288*.

Wilson, A. G. 'Entropy in urban and regional modelling: Retrospect and prospect', *Geographical Analysis* 42 (2010a), 364–94.

Wilson, A. G. *Knowledge Power: Interdisciplinary education for a complex world*. Abingdon: Routledge, 2010b.

Wilson, A. G. 'Knowledge power: Ambition and reach in a re-invented university'. In *Re-inventing the University*, edited by R. Munck and K. Mohrman, Chapter 3, 29–36. Dublin: Glasnevin Publishing, 2010c.

Wilson, A. G. *The Science of Cities and Regions: Lectures on mathematical model design*. London: Springer, 2012.

Wilson, A. G., ed. *Urban Modelling: Critical concepts in urban studies*. 5 volumes. Abingdon: Routledge, 2013.

Wilson, Alan, ed. *Global Dynamics: Approaches from complexity science*. Chichester: Wiley, 2016a.

Wilson, Alan, ed. *Approaches to Geo-Mathematical Modelling: New tools for complexity science*. Chichester: Wiley, 2016b.

Wilson, A. 'New roles for urban models: Planning for the long term', *Regional Studies, Regional Science*, 3(1) (2016): 48–57. doi: 10.1080/21681376.2015.1109474.

Wilson, A. 'Data – The new utility'. Supplement to *Prospect Magazine*, 'Data as infrastructure', ed. Duncan Weldon, October 2017. London: Prospect publishing company, 2017.

Wilson, A. 'Research and innovation in an ecosystem', *Journal of the Foundation for Science and Technology* (FST Journal) 22(3) (2018): 9–10.

Wilson, A. 'Epidemic models with geography: A new perspective on r-numbers', *ArXiv:2005.07673* [physics.soc-ph], 2020.

Wilson, A. G. and J. Dearden. 'Phase transitions and path dependence in urban evolution', *Journal of Geographical systems* 13(1) (2011a): 1–16.

Wilson, A. G. and J. Dearden. 'Tracking the evolution of regional DNA: The case of Chicago'. In *Understanding Population Trends and Processes*, edited by M. Clarke and J. C. H. Stillwell, Chapter 1, 209–22. Berlin: Springer, 2011b.

Wilson, A. G. and M. J. Oulton, 1983. 'The corner-shop to supermarket transition in retailing: The beginnings of empirical evidence', *Environment and Planning A* 15 (1983): 265–74.

Wilson, A. G. and C. M. Pownall. 'A new representation of the urban system for modelling and for the study of micro-level interdependence', *Area* 8 (1976): 256–64.

Wilson, A. G. and M. L. Senior. 'Some relationships between entropy maximising models, mathematical programming models and their duals', *Journal of Regional Science* 14 (1974): 207–15.Index

Page numbers in italics are figures and/or information in captions.

Index